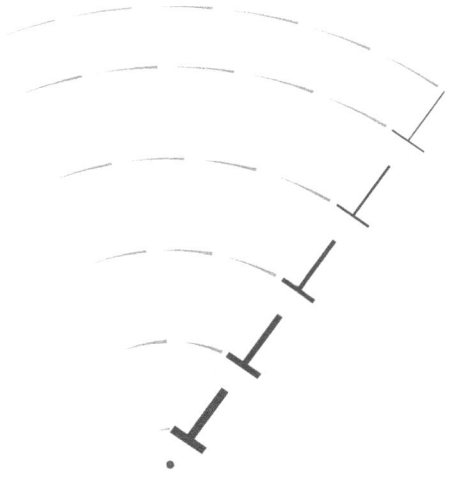

New Physics Framework

DAN S. CORRENTI

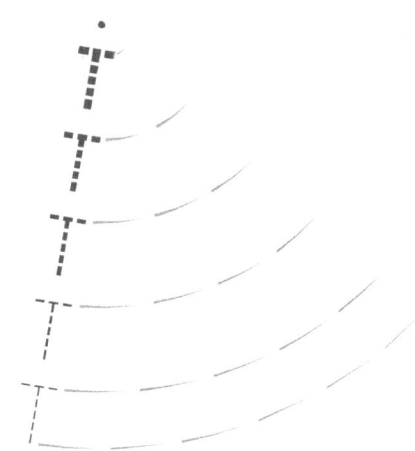

New Physics Framework

Oct. 15, 2019

Dan S. Correnti [a]

Abstract: Utilizing a classical perspective, the underlying physics and mechanisms of subatomic particles, atoms, photons, and the fundamental forces are examined and hypothesized. Such work is based on established models of the electric force and electron that were developed in a previous paper entitled: "Mechanisms explaining Coulomb's electric force & Lorentz's magnetic force from a classical perspective" in Elsevier's *Results in Physics* journal.[20]

Résumé: En utilisant une perspective classique, la physique et les mécanismes sous-jacents des particules subatomiques, des atomes, des photons et des forces fondamentales sont examinés et formulés comme hypothèses. Ce travail est basé sur des modèles établis de la force électrique et de l'électron développés dans un précédent article intitulé: «Mécanismes expliquant la force électrique de Coulomb et la force magnétique de Lorentz d'un point de vue classique» dans le journal Results in Physics de Elsevier.[20]

Keywords: subatomic particles, atoms, photons, fundamental forces

[a] dan@dcconsultingengineers.com
     630-858-6527

ISBN-10: 1475103204

ISBN-13: 978-1475103205

Drawings by Mrs. Plamena Yorgova

INTRODUCTION

The "Standard Model of Fundamental Particles and Interactions" is a generalization of quantum electrodynamics (QED) in which quantum chromodynamics (QCD) is modeled as similarly to QED as possible, although QCD is still found to be incomplete.[16] QED was initiated by Paul Dirac, who rewrote Erwin Schrödinger's differential wave equation for a charged particle in a relativistic form with spin operators. [21] However, the new form yielded implausible solutions such as negative energy states. To overcome these problems, the equation was reformulated and reinterpreted as a field equation.[17] Original or later versions of QED incorporated concepts such as quantization, uncertainty principle, creation and annihilation operators, renormalization, abstract operators and state space, virtual particles, and force mediators. The creation of quantum concepts such as these established a new mindset in physics – quantum logic.[18] While current QED is a probabilistic theory based on an abstract physical model and abstract mathematical formulation, it accurately predicts very precise measurements of various electromagnetic mechanisms and processes. Thus, QED is a testament to the impressive achievements of its developers in the science community.

Gravity and mass were not addressed in the model. Later, the Higgs boson was proposed and gained partial acceptance. Recently, this particle was "certified" as the mass-carrying particle. Nevertheless, gravitational theory (general relativity, which embodies mass and spacetime) is not yet understood at the quantum level. In addition, because gravity is much weaker than the other known fundamental forces (weak and strong nuclear forces and electromagnetism, which act over atomic scales), it cannot be reconciled with these forces in the Standard Model. To overcome these issues, cosmologists have proposed and developed other fundamental models (such as string theory), which incorporate the Standard Model. String theory, for example, accounts for the very weak gravitational force by introducing 10 or more spacetime dimensions, thereby lengthening the field over which gravity acts and offering an explanation for its weakness.

The founders of QED never intended to explain how and why a certain action occurred. Specifically, QED is mainly interested in predicting the probability of a certain action. Thus, mathematical objects and concepts were established in QED. Therefore, it is not necessary for QED to specify the makeup of charge, mass, and energy since it only uses the properties in its formulations to determine the probability of a certain action. Furthermore, QED is an extension of Dirac's probability wave equation that describes the probability of the occurrence of a potential event. It never describes how and why events occur or how particles are constructed because it does not necessitate this level of detail. Empirical "form factors" also allow QED to function without working at this level of detail. A form factor is a function that encapsulates the properties of a certain particle interaction without including all its underlying physics and instead provides the momentum dependence of suitable matrix elements. It is further measured experimentally to confirm or specify a theory.[19] For example, the fore-mentioned form factors allow the accurate determination of anomalous magnetic moment without understanding all the underlying physics.

The fore-mentioned models that constitute current physics framework, while being mathematically adept at describing numerous mechanisms and processes, are based mainly on quantum logic, without classical input; and because current models are mathematical rather than physical constructs, they cannot bridge large gaps that persist in our understanding of the physical world. For example, "What do electrons, protons, neutrons, and photons look like and how do they work"? Similar questions can be asked of fundamental forces whose current models are also mainly mathematical constructs. Both quantum and classical logic need to be utilized in order to overcome such deficiencies. For example, see the discussion below on Compton scattering.

The fact that scientists developed and grew quantum mechanics into one of the most precise science fields without the benefit of having physical models of sub-atomic particles and photons, to base their formulations on, makes their accomplishments more noteworthy. For example, if Maxwell did not have Faraday's models of electromagnetic field interactions to base his formulations on, his equations would not have been so specific and simplified. Thus, out of necessity, quantum mechanics had to be developed in an abstract and probabilistic manner, since physical models for it were not available.

Unfortunately, due this problem also, quantum mechanics was formulated and developed independently from gravitational theory. Thus, the two theories cannot be reconciled with each other since they are unrelated. Another concern is Dirac's equation, which is a basis for quantum mechanics, was formulated independently from Maxwell's equations. Since they both deal with electrons and photons, they should be related.

The results of this study aid in understanding the electron structure and its interactions by using classical mechanics while QED is required to provide the probability of particle behavior that results from blind interactions. This can be easily elucidated with an example. Specifically, with respect to Compton scattering, if the structures of the electron and photon as well as their relative orientations and positions at the point of interaction are known, then it is possible to accurately calculate the scattering direction of each object and also calculate the exchange of energy between the two objects. These variables are calculated by using classical mechanics. However, it is not possible to obtain the relative orientations and positions of the two objects, and thus they correspond to unknown parameters. Therefore, a probabilistic theory, such as QED, which accounts for factors including mass and momentum of objects is required to derive the probability of a given scattering angle or an exchange of energy between the objects.

In this study, an electron is found to be an individual self-contained field that does not emanate from a ubiquitous field. Whereas in 'quantum theory', ubiquitous EM fields, among other fields, are required as mathematical constructs such that the probabilistic behavior of interacting particles can be calculated, without knowing the makeup of such particles.

New physical models for the electron, proton, neutron, and photon are presented in New Physics Framework. If these models had been available to the physicists and mathematicians, when quantum mechanics was initiated and grew, the formulations for it would have been more specific and simplified. The formulations for quantum mechanics would have also been developed in conjunction with gravitational theory and Maxwell's equations; thus, interrelationships between the theories would have developed also.

<u>Synopsis for Mathematics used for the Electron Model:</u> Our studies have indicated that an electron can be viewed as an individual field (w/ particle & wave characteristics, i.e. double slit experiment) unto itself, which has electrical and magnetic properties. Hence, it should be possible to extract the physical model of the electron from the mathematics of Maxwell and Coulomb because the electron, itself, only consists of electric and magnetic interactions of its field components. These types of interactions are ruled by Coulomb and Maxwell's equations, thus manipulations of these equations should lead us to the electron model.[20]

This procedure is similar to that employed by Maxwell when he formulated his equations based on the electric/magnetic interaction models developed by Faraday, except in this case, the procedure is reversed. Coulomb/Maxwell's equations are first manipulated to isolate variables and formations that may be applicable to the field configurations and characteristics that fit an electron model. After several adjustments and re-trials, the electron model is found that gives its known characteristics and behavior. This trial and error procedure is required, because the electron structure cannot be viewed as Faraday could view and measure his macroscopic interactions.

The above process has been accomplished and is presented in "New Physics Framework". From this model, it is now possible to conceptually, understand the make-up of a photon, proton, neutron, and the four fundamental forces. Conceptually, understanding of other phenomena such as magnetic fields, electric force, dipole moment, heat, and energy is also possible.

TABLE OF CONTENTS

1. The Nascent Universe 1
2. Matter Creation – *Building an Electron* 2
    A. The Oscillating Electron Field 7
3. Electron Field Interactions with Other Fields 12
    A. Linearly aligned Electron Fields 12
    B. Electron Fields at an Angle 13
    C. Electron and Magnetic Fields 15
4. Matter Creation – *Building a Proton* 18
5. Electron's Magnetic Dipole Field 21
6. The Electron and Electromagnetic Radiation 22
7. Hydrogen 26
    A. Hydrogen molecule 28
    B. Hydrogen gas structure 30
    C. Fine Structure Examination 32
8. Gravity 35
    A. Microscopic Field 35
    B. Macroscopic Field 39
    C. Neutron Mass 41
    D. Mass/Energy 41
9. Nuclear Fusion and Other Elements 43
10. Hadrons, Leptons, Bosons and Neutrinos 46
11. Closing Examinations 47
    A. Energy 47
    B. Momentum 48
    C. Temperature 49
    D. Cosmology 49
        D.1. B_θ Field Structure 50
        D.2. CMB Radiation 50
        D.3. Galactic Gravity 54
Notes & References 57

1. The Nascent Universe

In the new framework, the universe originally consists of massless, twirling, oscillating "heat fibers" [photon fibers, See 7A] made of energy, oriented, and moving randomly in a vacuum. At this stage, the universe is devoid of matter and contains only an extremely dense mixture of fibers. Conditions and development in the nascent universe are detailed in Section 11, but first, the reader should become acquainted with the intervening sections.

Each heat (photon) fiber twirls about its origin, oscillates along its own axis at near-light speed, and reverses its velocity at its ends (see Figure 1). During one quarter of an oscillation cycle, starting from its origin, linear portions of the fiber are continuously transformed to perpendicular parts through the Lorentz length contraction, as shown in sequence "a" in Figure 1. Thus, a spread of fiber parts occurs in the direction perpendicular to that of outward movement.

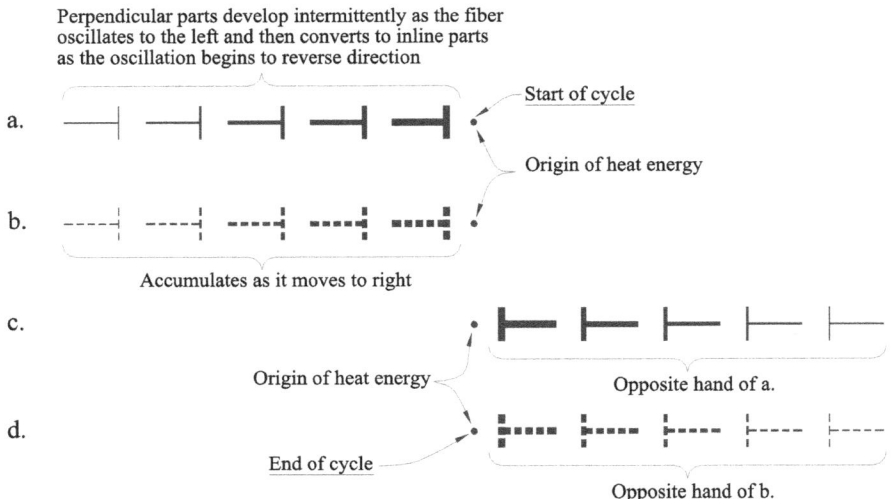

Figure 1 - Heat Fiber Oscillation

The intensity of the perpendicular parts is greatest near the origin, where the fiber volume is greatest, and gradually tapers toward the outer extreme of the fiber. Before the fiber retreats to the origin (sequence b), the perpendicular parts are returned to their original linear configurations by the reversal of direction. In sequence b, the fiber gains volume as it approaches the origin, and a new spread of perpendicular parts develops. As it crosses the origin, the fiber transforms into a spread of perpendicular parts on the opposite side (sequence c) and finally returns to its origin as in sequence b (sequence d), which completes one cycle of the oscillation.

2. Matter Creation – *Building an Electron*

The extremely dense mixture of heat (photon) fibers described above are twirling and oscillating and are randomly oriented and moving in a vacuum space. The fibers can be viewed as oscillating vectors possessing magnitude and direction. When two fibers meet, they interact strongly[23] due to the very dense mixture of them, and combine as vectors but remain as separate entities due to their perpendicular elements. When a double fiber randomly joins another single fiber, the triplet will orient in a direction weighted toward that of the doublet.[23] As more fibers join the ensemble, the fiber group becomes increasingly inclined to orient in the established direction. Within the framework, this process is generally referred to as the vector summation mechanism for reasons that will be explained later in this section.[15]

Fibers coexisting in the group avoid interference by adjusting their fiber spacing and orientation and aligning their origins and phases of oscillation.[23] Non-adjusting fibers such as those that do not orient into parallel planes do not join the group. Fibers rotating in a direction opposite from the group rotation also do not join. Again, this is part of the vector summation mechanism, which is later examined in this section.[15]

This mechanism ultimately creates a cylindrical field as the heat (photon) fibers combine with their origins aligning along a common axis. This structure is known as an electron (see Figure 2). The oscillating fibers and the cylindrical shape of the electron field are examined in more detail in Section 11 but the reader should first become familiar with the intervening sections. Each cross-sectional disk along the cylinder contains a radial heat (photon) fiber symmetrically twirling and oscillating around and across the longitudinal z-axis of the cylinder, respectively, as shown in the end view. The side view shows a snapshot of the field where the B-field disks in the central zone are "bunched-up" or compressed. Thus, the magnetic B-field is stronger in the central zone than in the outer zones.

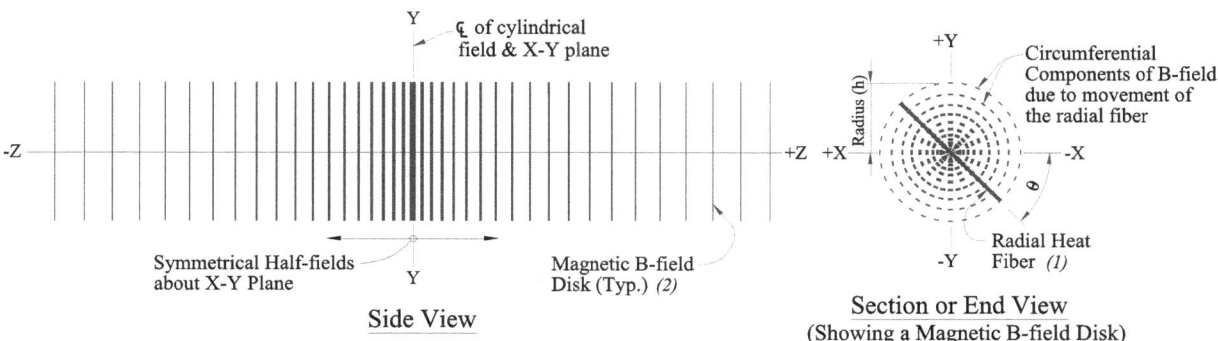

Side View

Section or End View
(Showing a Magnetic B-field Disk)

(1) Radial Fiber oscillates across and twirls about the Z-axis creating a spread of circumferential fibers - the B-field - perpendicular to and along the radial fiber as it passes; Circum. B-elements decrease in density as they occur further and further from the Z-axis. B-intensity increases with greater angular velocity of the fiber.

(2) Magnetic B-field Disks are represented as a group of individual parallel lines spaced further and further apart and less and less intense as they occur further and further from the centerline of the field in each direction along the Z-axis; In practice, the side view of the field would appear as a blur (due to the very close spacing of the disc lines) with decreasing intensity away from the centerline. The theoretical length and diameter of the field are infinity. The effective length and diameter are limited.

Figure 2 - The Electron

The cylindrical field on either side of the x–y plane oscillates along the z-axis.[15] As the two half-fields oscillate in or out along the z-axis, they rotate around the z-axis in opposite directions. Once both sides have fully compressed near the x–y plane, they reverse and move outward to their extremes, then reverse again to start another cycle. As both sides of the field move inward and outward, the whole field becomes more compressed and expanded, respectively. However, throughout the cycle, the central zone remains more compressed than the outer zones. Causes for field oscillation are examined later in Section 2. In this model, the impenetrable portion of the electron or proton field is a cylinder of length 2 fm and diameter 2 fm centered on the x–y origin, where the B-elements and fibers are heavily concentrated. The calculated "classical radius" of an electron is 2.82 fm and the implied radius of a proton (which is examined in Section 4) is 0.88 fm based on scattering tests; this data is only given for perspective since particles are not spheres in this study.

The radial fibers in the disks coexist because the disks are oriented in parallel. The oscillating and twirling motions of the radial heat fibers generate a circumferential magnetic B_θ field (hereafter referred to as the circumferential B-field, or simply "the B-field") in each disk (see Figures 1 and 2). The intensity of the B-field is highest near the z-axis (where the volume of the perpendicular components of the fibers is greater) and gradually decreases with increasing radial distance. The B-intensity also depends on the tangential speeds of the perpendicular components of the twirling fibers, which exert opposite effects on the field intensity distribution; this phenomenon is examined in Section 2A. The movement of the disks along the z-axis as the field oscillates is illustrated in Figure 3 below.

During contraction of the electron field, the disks in each half-field are propelled by the abovementioned vector summation of the radial fibers in the rear disks and the B-elements of adjacent forward disks. In this way, each half-cylindrical field is translated along the z-axis toward the x–y plane by simultaneous attraction between adjacent disks, and the disks "bunch up" at both sides of the x–y plane. Figure 3 shows a schematic of this process.

The rotational direction and intensity of the circumferential B-field are determined by the angular direction and angular velocity, respectively, of the twirling radial fibers in the disks. The greater the angular speed, the greater is the B-field intensity (see Figure 2). The intensity also depends on the density of the perpendicular components in the field, and on the oscillation range of the radial fibers. Because the innermost disks contain fibers with shorter oscillation ranges, the B-field is intensified there, as explained below.

As the components of the B-field disks advance toward the center, the B-elements accumulate and encourage further coupling of fibers. Each half-field becomes increasingly compressed as it approaches the x–y plane. Not only do the disks approach each other but their B-fields also intensify owing to the shortening oscillation ranges of the fibers. Thus, the overall compressed B-field is most intense around the x–y plane and gradually weakens at further distances. The decrease is more notable during outward movement of the half-fields.

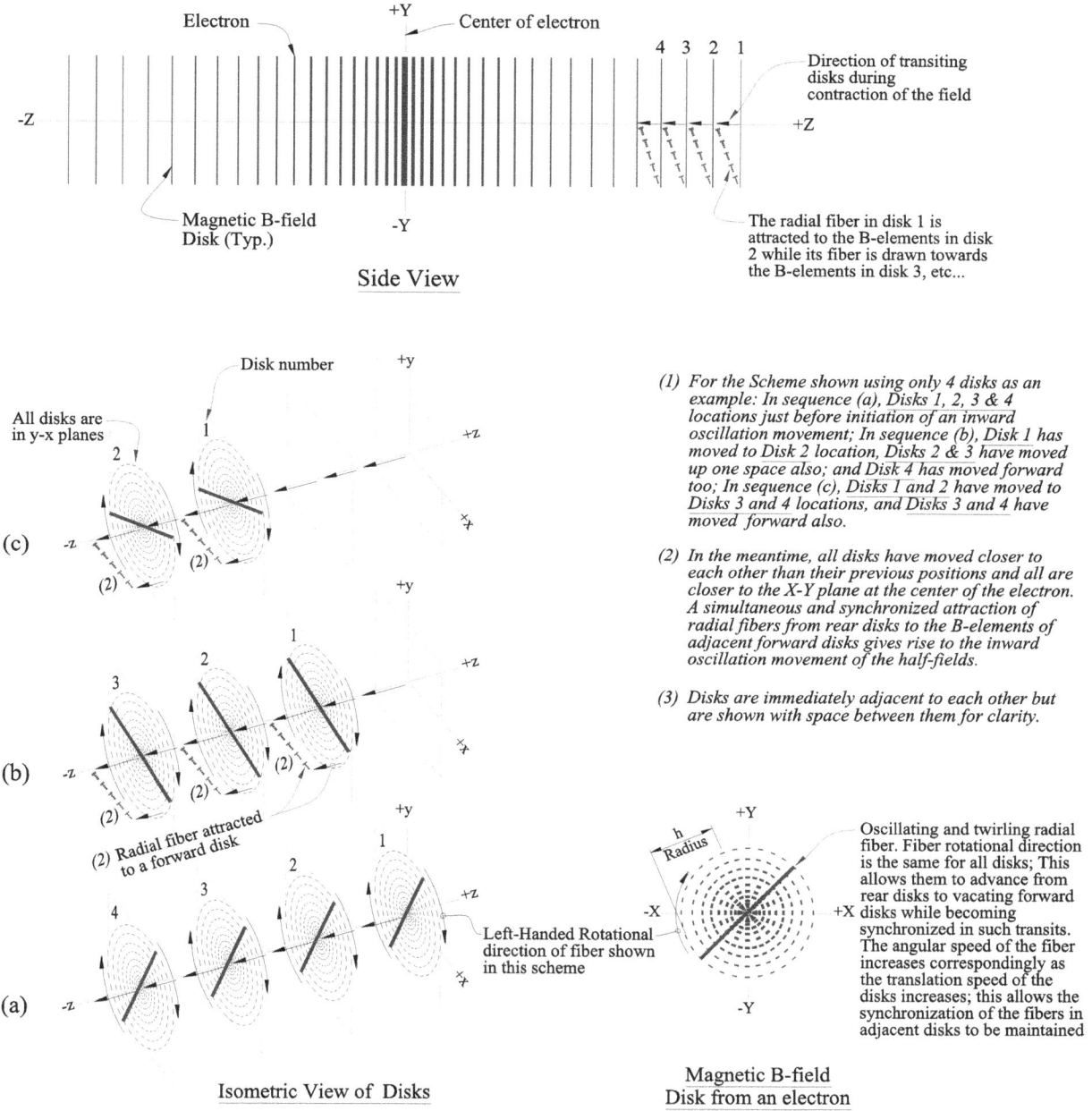

Side View

Isometric View of Disks

Magnetic B-field
Disk from an electron

(1) For the Scheme shown using only 4 disks as an example: In sequence (a), Disks 1, 2, 3 & 4 locations just before initiation of an inward oscillation movement; In sequence (b), Disk 1 has moved to Disk 2 location, Disks 2 & 3 have moved up one space also; and Disk 4 has moved forward too; In sequence (c), Disks 1 and 2 have moved to Disks 3 and 4 locations, and Disks 3 and 4 have moved forward also.

(2) In the meantime, all disks have moved closer to each other than their previous positions and all are closer to the X-Y plane at the center of the electron. A simultaneous and synchronized attraction of radial fibers from rear disks to the B-elements of adjacent forward disks gives rise to the inward oscillation movement of the half-fields.

(3) Disks are immediately adjacent to each other but are shown with space between them for clarity.

(1)(2)(3) Transitions of Outer 4 Disks During an Inward Oscillation Movement in Sequence (a) to (c)

Figure 3 - Oscillation Movement of Electron Field

The oscillatory movements and variations in the electron B-field are mathematically represented by the following Maxwell's Equations expressed in cylindrical coordinates, where J = 0 and v_z is the oscillation velocity of a half B-field along the z-axis.

$$\nabla \times B = \mu_o \varepsilon_o \frac{\partial E}{\partial t} \tag{1}$$

or specifically in cylindrical forms:

(1a) $\quad \dfrac{1}{h} \dfrac{\partial (h B_\theta)}{\partial h} = \mu_o \varepsilon_o v_z \dfrac{\partial E_z}{\partial z}$ $\quad$ *and* $\quad$ (1b) $\quad -\dfrac{\partial B_\theta}{\partial z} = \mu_o \varepsilon_o v_z \dfrac{\partial E_h}{\partial z}$

$$\nabla \times E = \frac{\partial B}{\partial t} \tag{2}$$

or specifically in cylindrical form: $\qquad\qquad$ *combining (1b) and (2a) gives*

(2a) $\quad \dfrac{\partial E_h}{\partial z} - \dfrac{\partial E_z}{\partial h} = v_z \dfrac{\partial B_\theta}{\partial z}$ $\qquad$ (2b) $\quad \dfrac{\partial E_z}{\partial h} = \left(1 + \dfrac{v_z^2}{c^2}\right) \dfrac{\partial E_h}{\partial z}$

These equations state that an accumulation of B-vectors or elements at any location or zone of the field increases E_z and E_h, the force potentials of the field in the z and radial directions, respectively, relative to another field. In other words, where B-vectors accumulate, they consequently attract the B-vectors of another field *(via their corresponding radial fibers)* moving in the same direction or repel fields with opposite direction. These mechanisms are mathematically expressed as absolute increases in E_z and E_h given by Equations 1 and 2.[15] In individual electron B-fields, the B-vectors *(via their corresponding radial fibers)* are drawn toward the electron center where other B-vectors accumulate. To put it in another way, the mobilization of E_z forces, which causes (for example) inward oscillation, are derived from the attraction of radial fibers in rear disks to the perpendicular components (B-elements) of radial fibers in adjacent forward disks (see Figure 3). Such types of movements of the B-field along the z-axis initiate and sustain oscillation of the electron B-field. The vector summation mechanism depicted in Figure 3 is described by Equation 1(a), while Equations 1(b) and 2(a) describe the geometric relationship between E_h, E_z, and v_z (see Equation 2(b) and Equation 4 in Subsection A). It should be noted that the mechanisms described by Equations 1(a) and 1(b) work together simultaneously, hand-in-hand, as follows: $\partial B_\theta / \partial z$ creates $\partial E_h / \partial z$, which causes $\partial(h B_\theta)/\partial h$, which creates $\partial E_z / \partial z$, which causes $\partial B_\theta / \partial z$, and the sequence of interactions repeat.

As the B-disks of each half-field transit toward the x–y plane, the effective diameter D of each half-cylindrical field gradually decreases to a minimum at the x–y plane. Although the electron field is illustrated as a uniform cylinder, the actual shape of each half-field is a curved truncated cone. The cones contact end-to-end at the x–y plane (in a subsequent section, we will show that D^2 is proportional to z^3). This shape arises from the increasing concentration of B-elements within the inner zone. The outer disks also approach and accumulate in the inner zone. The high concentration of perpendicular fiber components (B-elements) in the inner zone (particularly near the z-axis) attracts fibers in rear disks, whose oscillation ranges accordingly decrease. Thus, the diameter of each half-field is minimized at the x–y plane. The vector summation of this process is given by Equation 1(b). Although both innermost disks at the x–y plane possess the smallest diameter, these disks alone increase their diameter during the inward oscillation of their half-fields, in response to the positive net E_h force. All other disks experience a negative E_h type force that causes their diameters to decrease. As the half-fields oscillate outward, the disks return to their previous sizes, in accordance with Equation (1b). The disks of the half-fields remain "bunched-up" near the x–y plane but to a lesser extent; thus, the diameter of each half-field still gradually decreases from the outer to inner disks.

Because all fibers oscillate at near-light velocities during transit of the half-fields, their oscillation frequencies increase with decreasing oscillation range as their representative disks approach the x–y plane. Increasing frequencies are associated with increasing fiber energies. The decreasing disk diameter is accompanied by an exponential increase in B_θ intensity.

The right side of Equation 2 is negative (not shown) in Maxwell's Equations because the magnetic field generated by the induced emf and its current opposes the applied magnetic field. This effect is observed in unsymmetrical interactions among multiple fields, as depicted in Faraday's Law. In this case, Equation 2(a) is expressed in Cartesian coordinates as follows: B_θ is replaced with an applied B-field perpendicular to the z-axis of the electron; $\partial E_z/\partial h$ vanishes; E_h is replaced with an electrostatic potential perpendicular to both applied B-field and z-axis; and v_z denotes the translational velocity of the whole field rather than the oscillation velocity of a half-field. The Faraday equivalent of Equation 2(a) is written as

$$\frac{\partial E_{\perp B, \perp z}}{\partial z} = -v_z \frac{\partial B_{\perp z}}{\partial z} \qquad (2a')$$

In this study, the electron is modeled as a single symmetrical cylindrical field. Because its field interactions are purely internal and symmetrical, the right side of Equation 2 is unsigned and Equation 2(a) is expressed in the given cylindrical form. The electric force potential E induced by the B-field will be further examined in Subsection A. For now, we assume that E_z, the z-component of the electric force potential, proportionally varies in accordance with the variation in the B-field, as indicated in Equations 1(a) and indirectly in 2(b). If the change in B-field at a particular test location increases the absolute intensity of B_θ, then the net absolute change in E_z relative to an imaginary field at that test location proportionately increases. During interactions with other fields, the oscillation position of the half-field that exerts the maximum force on (or couples with) the external fields will control the interaction. This idea is elaborated in Subsection A.

The interaction reach of an electron field is greatest when the two half B-fields are fully extended. Figure 5 shows two overlapped electron fields with fully extended half-fields. The greatest interaction force occurs when the compressed inner zone (the region of high B-field strength) overlaps with another compressed field, such as an electron field. This overlap of compressed zones gives two electrons in close contact, yielding the maximum absolute change in B_θ, ($\partial(hB)/\partial h$ in Equation 1(a)) and therefore the highest absolute value of E_z. Both levels of interactions are embodied in Equations 1 and 2 and will be shown to underlie Coulomb's inverse distance square law in electrostatics.

As described above, the tendency of each half-cylindrical field in Figure 2 to simultaneously contract and move inward from both ends along the z-axis compresses the inner zone around the x–y plane. Correspondingly, each half B-field disk at the origin ($z = \pm 1$ fm) crosses the x–y plane at the center of the electron to the opposite side of the x–y plane. As these B-field disks with opposite rotational directions intercept in the space between $z = -1$ to $+1$ fm, they counteract each other, causing the inward movements of the cylindrical half-fields to reverse to outward movements. The vector summation that propels the outward movement is that depicted in Figure 3 but in the opposite direction, rather than gravitating toward the electron center, the disks and radial fibers are forced outward, consistent with Equation (1a). This outward movement is quickly resisted by the B-field disk components, especially those in the central zone, as the outward velocity slows and E_z between the disks decreases. This restraint gradually builds up until it overcomes the inertia of the outward expansion. The half-portions of the cylindrical field are again drawn toward the central zone and the cycle starts over.

As stated above, the forces responsible for halting the outward movement of the oscillating cylindrical field can be analyzed in terms of Maxwell equations based on vector summation of the B-field disks. The reverse action, in which the inward movement is halted by increasing outward force, is similarly treated but in a manner opposite to the inward-acting force. During inward oscillatory movements and as the fiber of the innermost disk of each half B-field initially draws to the innermost B-elements on the opposite side of the x–y plane, each fiber instantaneously resists the other by their opposing rotational directions, forcing both fibers to reverse their translational and rotational directions. Consequently, the disks in each half B-field decelerate and eventually stop, reverse their angular and translational direction, and traverse outward. As the inward-acting force gradually dominates, the cycle repeats.

Although the magnetic B-field is induced by the oscillating and twirling radial heat fiber in each disk, its intensity and direction are only measurable when the whole cylindrical field translates along the z-axis in either direction and the velocities of the disks sum to the whole field velocity. If the cylindrical field is stationary, the half-portions of the field inhabiting each side of the x–y plane traverse the z-axis but their B-fields are immeasurable because the angular and translational directions of each oscillating half-field repeatedly reverse. Because the strengths and movements of each half-field are equal and opposite, they cancel exactly.

A. The Oscillating Electron Field: As the half-portions of the symmetrical electron field oscillate between the center and extremity on both sides of the x–y plane, their component velocities vary along the direction of movement, the z-axis. Thus, the speed of the entire half-field and also its components vary along the z-axis. The speed variation along the z-axis is given by Equation (5), derived from Maxwell's Equations and Coulomb electrostatics. The derivation (which the reader is encouraged to follow) is based on the following arguments.

Although Coulomb forces are radially uniform from the center of an electron field, the derivation below considers only their effect along the z-axis. The electron field can oscillate and propagate only along the z-axis. For example, when an electron field enters the influence of another randomly oriented field, such as another electron field, both fields rotate such that their z-axes are aligned. This idea is examined further in Section 3B. Thus, the r^2 term (= $h^2 + z^2$ in Equation 5), where r denotes the distance between the charge centers of two particles (in this study, the distance between the centers of two electron B-fields), can be replaced with z^2 because the subject fields oscillate and interact along the z-axis. The velocity v_z depends on z alone.

Although h vanishes in Equation 5, it is required in its derivation because the B-field depends on both h and z. Moreover, by symmetry, the electric force associated with the B-field has only a z-component, but Maxwell's equations include the orthogonal E-components, which (unlike the B-distribution) are measurable for a static "charge." Therefore, effectively, Coulomb's force law for a non-translating electron B-field is one geometric form of Maxwell's Equations 1 and 2. Thus, the $\partial E_z/\partial z$ and $\partial E_z/\partial h$ components associated with the Coulomb force correspond to the $\partial E_z/\partial z$ and $\partial E_z/\partial h$ components in Maxwell's Equations 1 and 2, as indicated below in Equations (a) and (b). Equation (4) derives from the fact that Equation 2(a) is geometrically related to Equation 1(b), and Equation (b) follows. These relationships are consistent with Equation 2(b), which adjusts E_h by the Lorentz length contraction in the z-direction. It should be clarified that $\partial E_h/\partial z$ in Maxwell's equations is general and can relate to a Lorentz-type force such as Equation 2(a′) for example, whereas $\partial E_h/\partial z$ obtained from Coulomb's Equation solely relates to the Coulomb force and sums to zero.

The following derivation is based on the half B-field of a stationary electron that has extended from a compressed state along the +z-axis to a position where it can interact with another field. By the right-hand rule with the thumb pointing in the +z-direction, the circumferential direction of B_θ is positive in the clockwise direction. The positive z-axis points away from the viewer.

Derivation of the Electron's cylindrical B-field (CBF) Oscillation

Eq. 3 is obtained from Eq. (1a):   *Eq. 4 is obtained by combining Eq. (1b) & (2a):*

$$\frac{\partial E_z}{\partial z} = \frac{c^2}{v_z}\frac{1}{h}\frac{\partial(hB_\theta)}{\partial h} \quad (3) \qquad\qquad \frac{\partial E_z}{\partial h} = -\frac{c^2 + v_z^{\,2}}{v_z}\frac{\partial B_\theta}{\partial z} \quad (4)$$

v_z: variable oscillating velocity of a subject half B-field in the z-direction and as a function of z. When the half-field, on the +z side of the x-y plane, is moving away from the viewer in the +z direction, v_z is positive.

$$E = \frac{|q|}{4\pi\varepsilon_o r^2} \quad \text{Coulomb's Eq.} \quad \text{where } r^2 = z^2 + h^2 \quad \text{or} \quad r = \sqrt{z^2 + h^2}$$

q: e, Coulomb's electron charge constant – but in this study, a representative field constant of the electron's half B-field intensity and rotational direction.

E: Coulomb's electric force field – but in this study, the max. potential force exertion on another B-field, having a unit q, by the subject B-field at the instant when they are separated by a distance of r.

From Coulomb's Eq.:

$$E_z = \frac{|e|}{4\pi\varepsilon_o}\frac{z}{\left(z^2+h^2\right)^{3/2}}$$ from which $$\frac{\partial E_z}{\partial z} = \frac{|e|}{4\pi\varepsilon_o}\frac{h^2-2z^2}{\left(z^2+h^2\right)^{5/2}} = \frac{c^2}{v_z}\frac{1}{h}\frac{\partial\left(hB_\theta\right)}{\partial h}$$ from Eq. 3 gives: (a)

and $$\frac{\partial E_z}{\partial h} = \frac{|e|}{4\pi\varepsilon_o}\frac{-3hz}{\left(z^2+h^2\right)^{5/2}} = -\frac{c^2+v_z^{\,2}}{v_z}\frac{\partial B_\theta}{\partial z}$$ from Eq. 4 gives: (b)

From Eq. (a): $$hB_\theta = \frac{v_z}{c^2}\frac{|e|}{4\pi\varepsilon_o}\int\frac{\left(h^2-2z^2\right)h}{\left(z^2+h^2\right)^{5/2}}\partial h$$

and $$hB_\theta = -\frac{v_z}{c^2}\frac{|e|}{4\pi\varepsilon_o}\left[\frac{h^2-2z^2}{3\left(z^2+h^2\right)^{3/2}}+\frac{2}{3\left(z^2+h^2\right)^{1/2}}\right]$$ and thus, $$B_\theta = -\frac{v_z}{c^2}\frac{|e|}{4\pi\varepsilon_o}\frac{h}{\left(z^2+h^2\right)^{3/2}}$$ (c)

From Eq. (b): $$B_\theta = \frac{|e|}{4\pi\varepsilon_o}\int\frac{v_z}{c^2+v_z^2}\frac{3hz}{\left(z^2+h^2\right)^{5/2}}\partial z$$ gives: (d)

Equating (c) and (d): $$-\frac{v_z}{c^2}\frac{|e|}{4\pi\varepsilon_o}\frac{h}{\left(z^2+h^2\right)^{3/2}} = \frac{|e|}{4\pi\varepsilon_o}\int\frac{v_z}{c^2+v_z^2}\frac{3hz}{\left(z^2+h^2\right)^{5/2}}\partial z$$

and $$-\frac{v_z}{3c^2\left(z^2+h^2\right)^{3/2}} = \int\frac{v_z}{c^2+v_z^2}\frac{z}{\left(z^2+h^2\right)^{5/2}}\partial z$$

Differentiating with respect to z: $$-v_z\left(-\frac{z}{c^2\left(z^2+h^2\right)^{5/2}}\right)-\frac{\partial v_z/\partial z}{3c^2\left(z^2+h^2\right)^{3/2}} = \frac{v_z}{c^2+v_z^2}\frac{z}{\left(z^2+h^2\right)^{5/2}}$$

and $$\left(\frac{zv_z}{c^2\left(z^2+h^2\right)^{5/2}}\right)-\left(\frac{\partial v_z/\partial z}{3c^2\left(z^2+h^2\right)^{3/2}}\right) = \frac{zv_z}{\left(c^2+v_z^2\right)\left(z^2+h^2\right)^{5/2}}$$

and $$\frac{\partial v_z/\partial z}{v_z} = \frac{3z}{z^2+h^2}\left[1-\frac{1}{\left(1+\frac{v_z^2}{c^2}\right)}\right]$$ and $$\frac{\partial v_z/\partial z}{v_z}\left(1+\frac{c^2}{v_z^2}\right) = \frac{3z}{z^2+h^2}$$

Integrating with respect to z: $$\int\frac{\partial v_z}{\partial z}\left(\frac{1}{v_z}+\frac{c^2}{v_z^3}\right)\partial z = \int\left(\frac{1}{v_z}+\frac{c^2}{v_z^3}\right)\partial v_z = \int\frac{3z}{z^2+h^2}\partial z$$

and $$\ln|v_z|-\frac{c^2}{2v_z^2} = \frac{3}{2}\ln|h^2+z^2|$$ (5)

As explained above, $h^2 + z^2$ in Equation 5 reduces to z^2. The velocity as an exponential function of z, given by Equation (6), is plotted in Figure 4. This figure assumes that each half-field is extended. As each half-field oscillates inward and contracts, its innermost disk crosses to the opposite side of the y-axis. As the disks meet in the crossover zone (z = −1 to +1 fm in this study), each is repelled by the opposing rotational direction of the other. Consequently, the inward velocities of the intercepting disks are substantially reduced, causing them to halt and reverse outward.[1] Figure 4 also shows the crossover zone of the innermost disks. Each dashed line in the figure traces the movement and velocity of the innermost disk of the half-fields as they move from an expanded phase to a contracted phase and *vice-versa*.

$$\ln|v_z| - \frac{c^2}{2v_z^2} = 3\ln|z| \qquad (6)$$

The distribution of the velocities (v_z) along the z-axis in Figure 4 occurs when the half-field is extended and has maximum potential to interact with another field. For example, if the center of another electron field is located at z = 1 nm, the electron half-field extends outward to this point and exerts its maximum force at this instant. Its velocity v_z then varies between 0.099c and 0.212c, as shown in the graph. The velocity is traced relative to the origin of the electron and in most cases, the origin is also moving—even in static interactions.

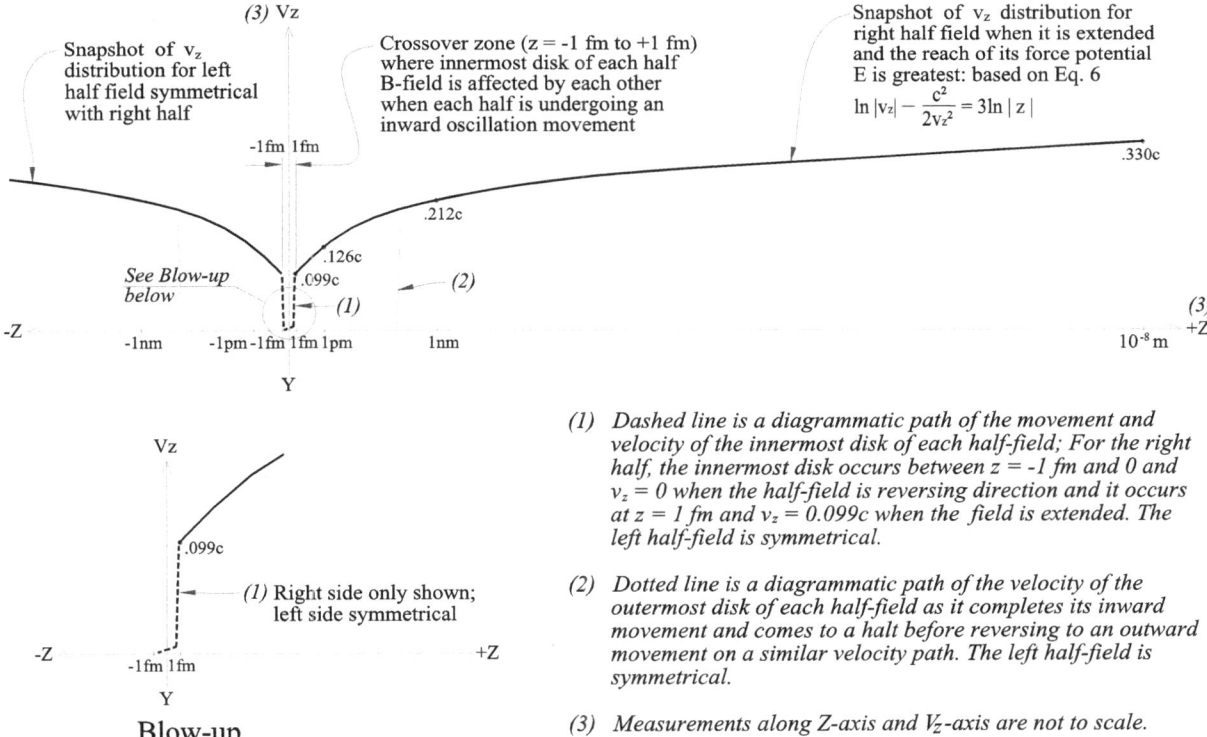

Figure 4 - Oscillation Velocities of Electron B-Field

The varying B-field intensities along the z-axis of the electron can be obtained at the instant of maximum force potential E_z at a specified location along the z-axis. Considering the above conditions for another electron field located at z = 1 nm, the velocities between z = 1 fm and 1 nm vary from 0.099c to 0.212c. Specifying a location between z = 1 fm and 1 nm, v_z is found from Equation 6; subsequently, the effective B_θ is found from Equation (c). This value of B_θ sums the B_θ increments from theoretical infinity to h and z, representing the B_θ strength along the continuum.

Differentiating Equation (c) with respect to h (i.e., calculating $\partial B_\theta/\partial h$) and equating it to zero, the location of maximum B_θ intensity, within a disk, h_{max} is given as 0.707z [14]. The maximum B_θ (calculated from Equation (c)) is then:

$$max.\, B_\theta \;\, of \;\, a \;\, disk \;\, at \;\, z = \frac{v_z}{c^2}\frac{e}{4\pi\varepsilon_0}\frac{0.39}{z^2} \;\; (c') \;\; \& \;\; eff.\, avg.\, B_\theta \;\, of \;\, a \;\, disk \;\, at \;\, z = \frac{max.\, B_\theta}{3} \;\; (c'')$$

Dividing the result by three in Equation (c'') gives the effective average B_θ intensity of an effective disk diameter (D_{eff}) as 9 × h_{max}, considerably smaller than the actual disk diameter (obtained in a subsequent section). Integrating Equation (c) with respect to h and summing the B_θ values from h = 0 to h = 9 × h_{max}, the summed B_θ equate to approximately 85% of B_θ of an actual disk. The average intensity of B_θ outside the effective disk is approximately 3% of the maximum but extends over to a much greater range. Although the average B_θ over the effective disk diameter adequately represents the average B_θ intensity of the disk, it is conservative when calculating interactions with other fields.

In this study, ±1 fm is the minimum distance from the y-axis at which the half-fields can begin transiting to the opposite sides of the axis. Inserting z = 1 fm into Equation 6, and inserting the resulting v_z into Equation (c') yields the effective B_θ intensity for each half-field in Equation (c''). This value equals the sum of B_θ increments along the z-axis of a half-field; that is, the strength of a full half-field giving rise to the Coulomb force potential of the electron B-field, obtained by "e" (the electron charge constant, but is a field constant in this study). This can be seen from Equations (a) and (b) in which the summation of B_θ increments is associated with summation of E_z increments, and thus E_z, the Coulomb's force potential. The distribution of B-field disks along the z-axis (depicted in Figure 2) is mathematically described by Equation (c), which is derived from Equation (a).

Also noteworthy in Equation (c) is that the circumferential direction of B_θ is negative (counterclockwise) when each electron half-field moves outward from its center. The direction of net B_θ of the whole propagating electron field is that of the forward half-field moving outward from the center of a stationary electron. By inserting an absolute value of "e" in the derivation above, this direction was determined as counterclockwise. In this study, the correct sign of the electron B-field constant "e," ensuring that the electron field abides by the left-hand rule, is positive. Thus, the "e-field" constant of the proton (see Section 4) is negative, indicating that the clockwise circumferential direction of its B-field follows the right-hand rule.

3. Electron Field Interactions with Other Fields:

The whole cylindrical electron field can propagate in either direction along the z-axis only when propelled by the vector summation mechanism of its disk components responding to (for example) the disk components of another field. This mechanism is similar to that of Figure 3, where the vector summation mechanism causes the disks/fibers of the half-fields to move toward the x–y plane during an inward oscillatory movement. The net direction of the circumferential B-field is that of the radial fiber rotation, while sequencing. A stationary cylindrical electron field can be mobilized along the z-axis by overlapping its field with another electron field (Figures 5 and 6) or with an applied perpendicular moving magnetic field (Figure 7). These mechanisms are further explored in subsections A, B, and C below.

Throughout this study, we adopt the conventional sign convention specified by the right-hand rule; the right-hand fingers curl in the direction of the B-field when a positive charge translates along the direction of the extended right thumb. Thus, the B-field and translation of a moving electron field follow an equivalent left-hand rule. To ensure this rule for the electron, the radial heat fibers of the cylindrical field must have a net "left-handed" rotation. Because the net sum of the velocities of both half-fields is in a single direction, the corresponding net angular direction of the fibers and that of the B-vectors are left-handed.

A. Linearly aligned Electron Fields: Figure 5 shows two overlapping cylindrical electron fields. Each field is longitudinally aligned along the z-axis. As both fields approach or retract from each other along the z-axis, their B-fields oppose each other because the rotation of each follows the left hand rule. If both fields traverse the z-axis in the same direction, the B-fields remain opposed owing to the variation in their oscillating directions.

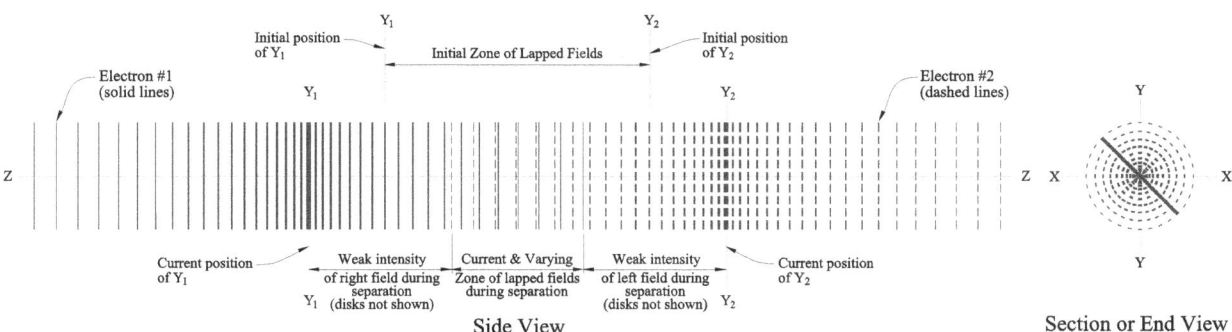

Figure 5 - Snapshot of Interaction of Two Inline Electron Fields

Thus, the B-fields of two oscillating electrons, whether at rest or translating along the z-axis, always oppose each other. By the same mechanism that causes the cylindrical field of individual electrons to oscillate and compress, namely vector summation of the B-field disk components, the opposing B-fields force the electron fields apart. Here vector summation causes the B-field vectors and the corresponding radial fibers in the overlapped disks to separate and retract from each other, rather than accumulate as in case of individual electron half-fields.

As the inner halves of both electron fields separate, their individual disks must pass each other. During this movement, the radial fibers comprising the disks, which are responsible for the B-fields, must either pass through or individually interchange. For example, consider two interacting electrons, 1 and 2. A fiber in a disk of electron 1 can detach and join a disk of electron 2 during the interaction and *vice-versa*. Fibers not interacting with fibers from the other field pass through unimpeded.

The varying B-field and speed of each electron half-field during the interaction shown in Figure 5 can be calculated similarly to the velocities and magnetic strength B of individual oscillating electrons by using the same formulas. As the half-fields separate, the intensities within the overlapped zone, and thus the repulsive force gradually decrease. To obtain v_z of the field of electron 1 at the origin of electron 2 (Y_2), the distance z between Y_1 and Y_2 is inserted into Equation 6. Inserting z and v_z into Equation (c′) and adjusting by Equation (c″) yields the effective B_θ, the sum of the B_θ increments from theoretical infinity to the calculation point. To obtain v_z of the field of electron 2 at its origin, z = 1 fm is inserted into Equation 6; v_z is then inserted into Equation (c′), in which B_θ is computed from Equation (c″). The product of the two calculated B_θ values is almost linearly proportional to the Coulomb force and represents vector summation (in this case, among the opposing disks' B-vectors and radial fibers) in the overlapped zone. The calculated v_z are the velocities of the respective fields at the calculation point relative to their origins.

As the electrons separate from one another, they translate in opposite directions. During translation, the B-field vectors and radial fibers of each electron field sequentially advance in a left-handed fashion as described above. When an electron translates, both halves of the field advance in the same direction instead of oscillating in opposite directions with zero net displacement, as occurs in the case of a stationary electron field. The net varying velocity in each half-field is obtained by adding its varying oscillation velocity to the overall translation velocity. Thus, at any instant, the velocity in each half differs by an amount equal to the translation velocity.

B. Electron Fields at an Angle: Figure 6 shows two overlapping electron fields oriented at an angle φ. The vectors shown at the four corners A to D of the overlapped zone are vertical B-vector components tangential to the B-field. Vectors 1 and 2 arise from electrons 1 and 2, respectively. Their directions are determined from the left-hand rule applied to the electrons oscillating in the directions shown. The illustrated vectors are representative of all tangential perimeter B-vectors (having a vertical component) along each successive incremental perimeter of each disk in the overlap zone.

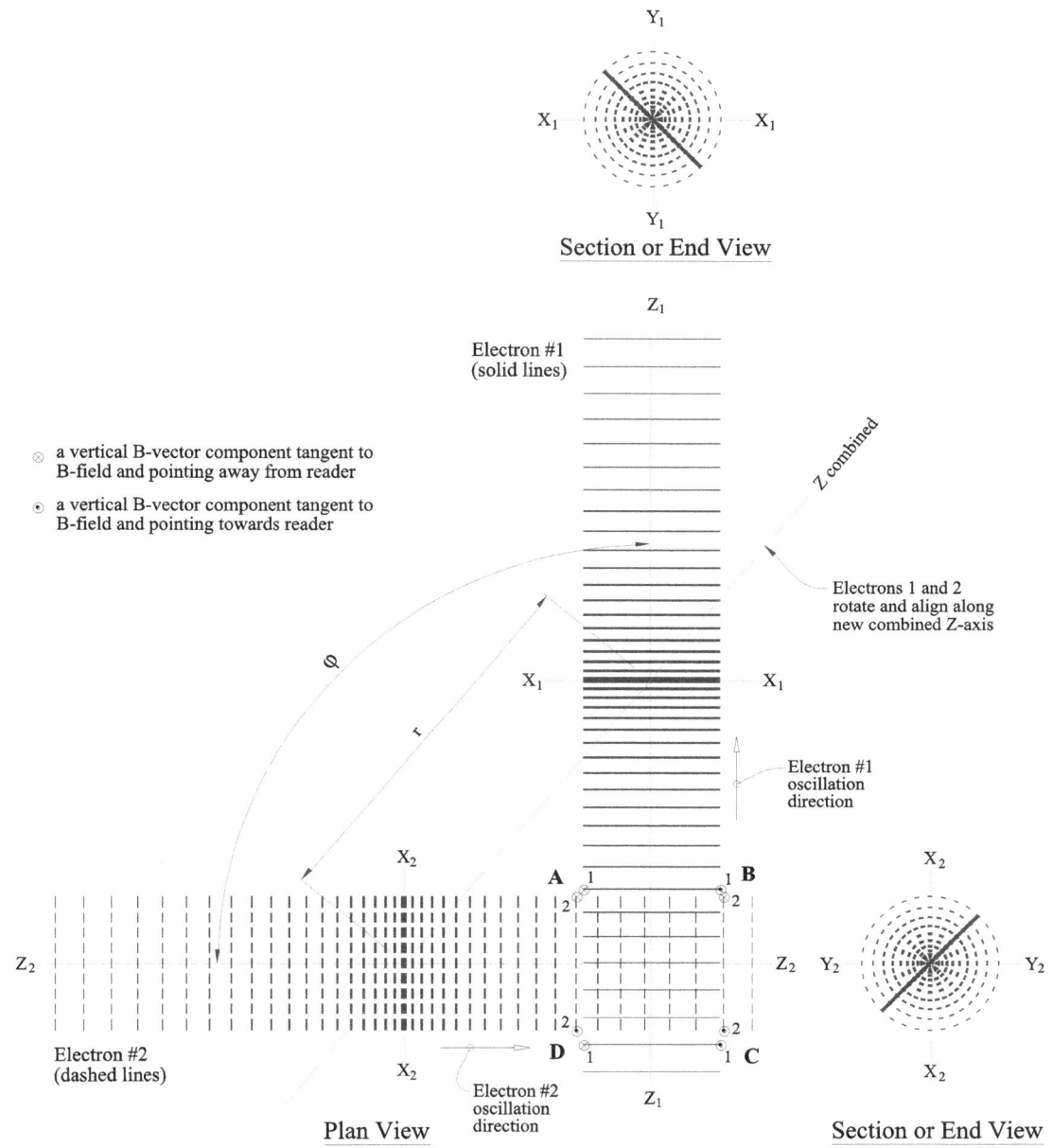

Figure 6 - Interaction of Two Electron Fields

For illustrative purposes, we assume that each pair of vectors at the corners has equal strength. In this case, vectors 1 and 2 at corners A and C point in the same direction and the effective strength at these corners is doubled. These vectors can also receive and accumulate other vectors, further increasing the strength at corners A and C. Conversely, vectors 1 and 2 at corners B and D oppose each other and their strengths cancel each other out. These vector summations cause the longitudinal axes of the two B-fields at each corner to rotate and combine. The new diagonal axis is determined by the direction of the thumb when curling the left hand fingers from corner A to corner C. Throughout this rotation, vectors D-1 and B-2 shift to corner A and accumulate there, while vectors B-1 and D-2 accumulate at corner C.

As the combined B-field at the corner forms and reorients, the corners and remainders of each electron field, whose disks are linked to the corner, rotate until both fields align along the new combined axis shown in Figure 6.[2] At this point, both electron fields are overlapped as shown in Figure 5 and interact as depicted therein.

The direction of oscillation of each electron field varies. If the directions are opposite to those of Figure 6, the axis of the combined B-field is rotated 180° from the axis shown in that figure. In both cases, the angle φ increases until the axis of each field aligns along the combined z-axis of Figure 6.

If both electron fields oscillate either outward or inward, the B-vectors accumulate at corners B and D and cancel at corners A and C. In this case, the combined B-field at the corner is oriented 90° from the previous case, where the oscillation directions are opposite. Thus, the summation of the B-vectors at the corners causes a decrease in φ. At the start of this process, the B-field vectors of the two electron fields begin to interfere, prohibiting rotational movement. The standstill continues until, by their natural cycles, the two half-fields move in opposite directions, thereby meeting the criterion of the first case.

Thus in all cases, the angle φ between the electron fields increases to 180°, where the fields are rotationally stable. From this point onward, the fields interact as described in Figure 5, although simultaneous rotational and translational movements are possible. The distance r between the centers of the electron fields is the separation distance used in Coulomb's electric force law, and is valid for virtually any relative angular orientation between electron fields.

C. Electron and Magnetic Fields: Figure 7 shows an external magnetic field acting on a translating electron. The longitudinal z-axis of the electron is perpendicular to the applied magnetic field. The directions of the tangential B-vector components 1, 2, 3, and 4 shown in the end view follow the left-hand rule for an electron field translating in the +z direction. These vectors graphically represent the net sum of all tangential perimeter B-vectors (components parallel to the principal axis) over each incremental perimeter in each disk of the whole B-field (i.e., both halves) and are drawn as tangents to the principal axes in Figure 7. The B-vector tangents on the positive y side of the x-axis are oriented parallel to the applied magnetic field, and are referred to as 1x-tangents, while those on the negative y side of the x-axis are oriented anti-parallel to the applied magnetic field, and are called 3x-tangents.

Then, for an electron moving in the +z direction, the applied B-field gravitates toward the 1x-tangents of the B-vectors while opposing the 3x-tangents, causing the whole electron field to shift in the +y direction. This tendency results from the vector summation mechanism previously described (Equation 1b). However, the distributions of the B-vectors in each half-field, if examined closely, are consistent with Lenz's Law, which dictates the direction of movement of the electron field. Lenz's Law directs the field in the −y direction rather than the +y direction, and the right-hand side of Maxwell's Equation 2 becomes negative. The force causing the downward movement is given by the magnetic component of Lorentz's force equation, calculated from Equation 2(a′).

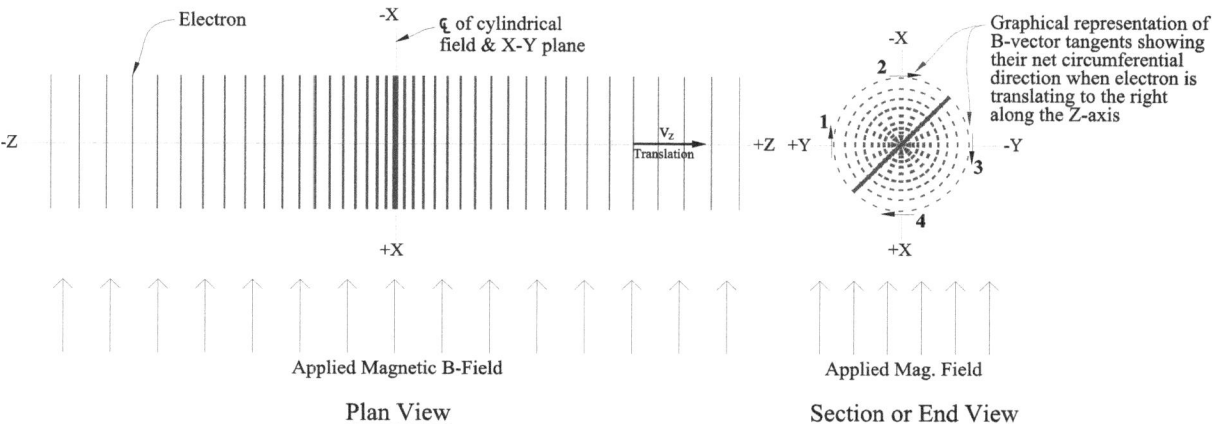

Figure 7 - Interaction of Magnetic Field and Electron Field

For a physical interpretation of Lenz's Law and the Lorentz force, we first examine the behavior of a stationary electron subject to an applied magnetic field (see Figure 7). When both half-fields of a stationary electron oscillate toward the x–y plane, the 1x-tangents of the half-field on the −z side are aligned with the applied magnetic field, while the 3x-tangents oppose it. In the +z half-field, the reverse is true; the 3x-tangents are aligned with the applied magnetic field, while the 1x-tangents oppose it. When both half-fields oscillate away from the x–y plane, the 1x- and 3x-tangents in both half-field reverse their directions. Therefore, in both cases, the pairs of 1x- and 3x-tangents on opposite sides of the x–y plane are equal and opposite and thus cancel each other. Because the net circumferential direction of the B-field is zero, B-vectors and their opposing counterparts cannot accumulate with and oppose the applied field in the +y zone and −y zone, respectively, on a net basis as occurs in a moving electron. Consequently, the electron cannot move in the perpendicular direction. Although the net force on the field in the y-direction is zero, the electron field is torqued in the y–z plane by the equal and opposite distribution of the B-vector tangents symmetrically located about the x–y plane in the half-fields. For example, if both half-fields oscillate outward, the electron rotates counterclockwise in the y–z plane. When the oscillation direction reverses, the field rotates clockwise and returns to its previous position. Thus, the stationary electron field cycles in the y–z plane. The centerline of its cycling range is the y-axis, and its mean longitudinal direction is the z-axis.

An electron field translating in the +z-direction, as shown in Figure 7, also rotates in the y–z plane, but the centerline of its cycling range rotates clockwise from the y-axis. Its mean longitudinal direction is rotated an equal amount clockwise from the z-axis. Thus, the forward and rear half-fields are inclined below and above the z-axis, respectively. Because the forward half-field can translate only along the z-axis, its downward inclination predetermines its downward movement in the −y-direction, as required by Lenz's Law. The z-axis continually rotates clockwise as the field shifts downward. Without a downward inclination of the field, the z-axis would shift in the +y-direction owing to B-vector summation under an applied field as previously described. In this case, however, the electron field experiences an upward force that is restrained by the declination of its z-axis. This causes the electron to move in a downward curved trajectory at a radius and velocity that define the trajectory.[3]

The translating field moves downward because the distributions of the 1x- and 3x-tangents of the B-vectors differ between the forward and rear half-fields. Under the applied field, the accumulation of 1x-tangents and opposition of 3x-tangents is greater in the rear than in the forward half-field. Thus, the net upward force is greater in the rear half-field. The rear half-field "sweeps in" (via the vector summation mechanism) more of the applied field when oscillating inward along the z-axis than when oscillating outward, although the effect is greater than in the forward half-field in all linear motions. The reasons are twofold: 1) the velocity of the rear half-field relative to the magnetic field is maximized during inward movement, and 2) the rear-field translates at its greatest relative velocity when its B-field compresses. During the compression phase, the spatial gradient of B_θ ($\partial B_\theta/\partial z$) is increased, enhancing the accumulation of the applied field and thus the perpendicular component of E.

Although the maximum relative velocity of the forward half-field oscillating outward equals that of the rear field oscillating inward, the B-field of the forward field decompresses during this phase. In addition, the forward half-field becomes less compressed than the rear field during inward movement. Thus, despite the velocity equivalence, the forward half-field "sweeps in" less of the applied field.

The abovementioned downward shift may also occur in a stationary electron if the applied magnetic field shown in Figure 7 moves in the −z direction. As the electron field oscillates along the z-axis, the rear half-field "sweeps in" more of the moving magnetic field than the forward half-field, for reasons discussed above. The effect on the electron is identical: a net upward force acting on its rear half-field causes the whole field to decline clockwise. Consequently, the electron field is shifted downward and rotated clockwise.

The longitudinal z-axis of the electron field is always oriented perpendicular to an applied uniform magnetic field, as shown in Figure 7. An electron intercepting such a field will rotate in its x–z plane to satisfy this condition, if no other field exerts an influence. If the applied magnetic field is moving relative to the stationary electron, the electron field will also rotate in its y-z plane. It also undergoes a y-directional shift, which is perpendicular to both the applied field and its propagation direction.

The alignment of the z-axis of the stationary electron perpendicular to the applied field is explained by the vector summation mechanism, in which the 1x-tangents and 3x-tangents of the B-vectors of the cylindrical field reorient parallel to the applied magnetic field and then accumulate or oppose the field. Consequently, the disk components (and hence the disks) of the cylindrical field shift in sequence, and the whole field rotates.

The downward Lorentz force (equal and opposite to the resultant upward force acting on the rear half-field), calculated from Equation 2(a′), results from vector summation of the 1x- and 3x-tangents of the B_θ field with the applied B-field, as configured in Figure 7. The summation of all B_θ tangent increments is given by the B_θ intensity of the whole field, obtained from Equation (c) with z = 0, h = 1 fm and v_z satisfying $v_z^2 \ll c^{2[10]}$.[11] In this case, v_z is the translation velocity of the whole field, not the oscillation velocities of the half-fields (which exert no effect on the translation component of B_θ). Furthermore, z = 0 because the translational, rather than the oscillatory, B_θ increments are summed. The vector summation represents the product of the obtained B_θ intensity and the applied B-field intensity, and is proportional to the Lorentz force (Equation 2a′). The Lorentz force is evaluated as follows:

$$F_L = -B_\perp \left[B_\theta \, \frac{4\pi}{\mu_o} (1\,fm)^2 \right]$$

$$\text{(equiv. to ev)}$$

(7)

The product of e and v_z in the Lorentz force equation is replaced by an equivalent term calculated from Equation (c) by using the above-specified parameters. Note that B_θ assumes h = 1 fm. Thus, the area unit in Equation 7 is $(1 \text{ fm})^2$, which equals 1×10^{-30} m^2. If instead B_θ assumed h = 1 am, the area unit in Equation 7 would be (1 am),^{2} but the value of F_L would not change.

As previously stated, the Lorentz force is equal and opposite to the resultant upward force acting on the rear half of the electron field. This force is attributable to the denser B-field of the rear half-field during an inward oscillation movement as previously examined; thus, the mean field in Figure 7 inclines downward. Combined with the forward velocity of the whole field, this declination creates an equal downward force component—the Lorentz force—on the whole electron field. The field spirals clockwise due to the downward force component.

4. Matter Creation – *Building a Proton*

As electrons are formed, space is densely filled with their fields. These randomly moving fields repel each other, however, some are forced into close parallel alignment, as shown in Figure 8. Both fields in Figure 8(a) are shown in their expanded positions. When both fields oscillate in phase (as in Figure 8(a)), their B-vectors oppose in the y-direction where parts of their fields overlap, and the fields repel each other in the x-direction.

In most cases, the fields will oscillate out of phase with each other, and the aligning fields may attract instead of repel. For example, consider that the half-fields of electron 1 are compressed near the x–y plane and beginning to oscillate outward, while those of electron 2 are expanded and beginning to oscillate inward (Figure 8(b)). In this case, the two fields are ½ cycle out of phase. The directions of the B-vectors in the right half-fields (relative to the x–y plane) are governed by the left-hand rule for electron fields, and are shown in the end view of Figure 8(b). As indicated in the figure, the B-vectors of both fields align in the same direction, having Y-components (called tangents $B_{\theta(2Y)}$ and $B_{\theta(4Y)}$) in the overlapping field zone.

Because the vector summation mechanism [Equations 1(b) and 2(a) and similar to the force calculation of Equation 7] applies in the overlapped zone, the two fields approach each other along the x-direction and merge into a single field. Because half-field 1 is contracted, all of its B-vectors, and thus its B-field intensity, are found within a narrow zone, and the spatial rate of change in B_θ ($\partial B_\theta / \partial z$) is high. In the expanded half B-field 2, which is more dispersed along the z-axis, $\partial B_\theta / \partial z$ is lower. The enhanced $\partial B_\theta / \partial z$ of half-field 1 in the overlapped zone attracts the half-field 2 (Equations 1b and 2a). Consequently, the B-vectors of half-field 2 reverse and align with the B-vectors of half-field 1 as the two fields merge. During the merging process, half-field 2 overcomes the outward oscillatory movement of half-field 1 and completes its inward motion. This also results from the vector summation mechanism (Equation 1a) that propels the less dense part of the merging fields toward the denser part near the electron center. The disks in the right combined half-field are now more closely spaced.

Both half-fields on the left side of the x–y plane undergo the same interaction but are oppositely handed. Due to multiple interactions of these types *[see Note 4 for elaboration on such interactions]*, the electron fields merge into a single field—the proton—with a common z-axis and with clockwise circumferential direction of B-field vectors in each of its half-fields (based on the right-hand rule) as they oscillate inward and outward. As the proton field propagates along its z-axis, its summed B-vectors also rotate clockwise. By contrast, the electron field follows the left-hand rule and rotates counterclockwise during oscillation and propagation. Because the circumferential directions of their B-field vectors are opposite, the proton and electron fields attract each other.[4] The mode of this interaction is identical (but opposite in direction) to that demonstrated in Figures 5 and 6 for two repelling electron fields.

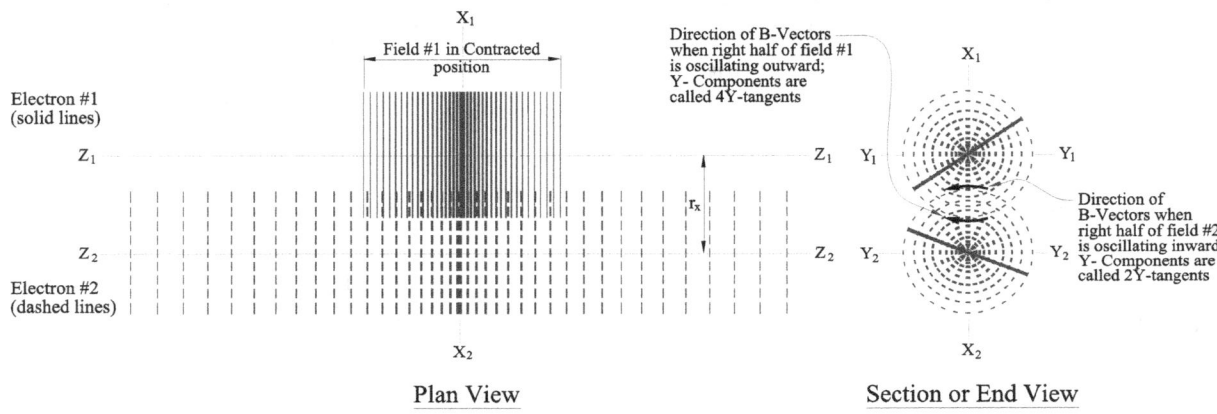

(b) Contracted/Expanded Positions out of Phase

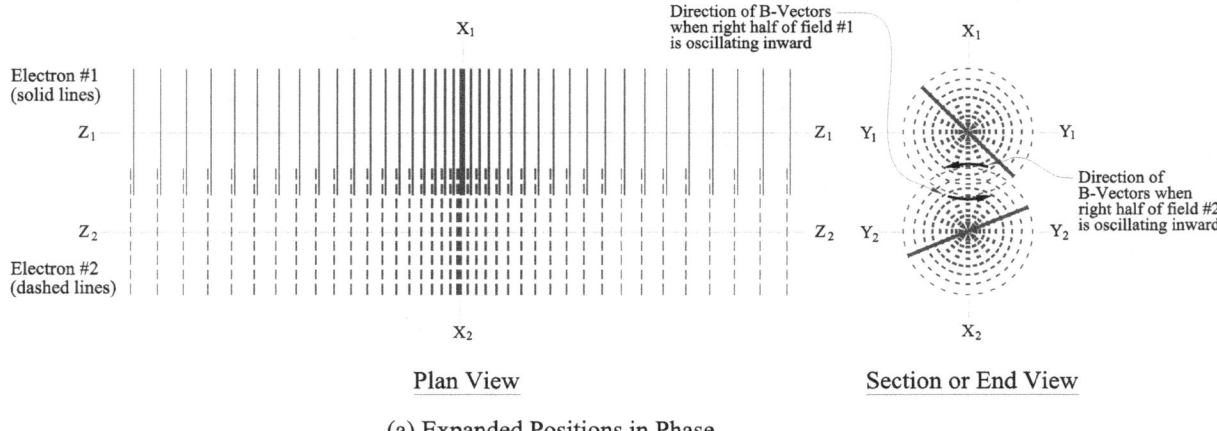

(a) Expanded Positions in Phase

Figure 8 - Interaction of Two Parallel Side by Side Electron Fields

The side-contacting electron fields in Figure 8(b) connect through an attractive force F_L, given by Equation 8 below. F_L is calculated similarly to Equation 7 in the preceding section, but is adapted for two interacting B_θ fields rather than a single B_θ field interacting with an applied field. $B_{\theta(2Y)}$ *(equivalent to B_θ in Equation 7)* is the magnetic component of field 2. The applied perpendicular B-field in Equation 7 is replaced with $B_{\theta(4Y)}$ of field 1, calculated along the z-axis of field 2. For details of these symbols, see the footnotes to Equation (8).

$$\overset{(3)}{F_L} = \left(\underset{\substack{\text{Field \#1} \\ \text{(eq. to } B_\perp)}}{\overset{(1)}{B_{\theta(4y)}}} \right) \times \left(\underset{\substack{\text{Field \#2} \\ \text{(equiv. to ev)}}}{\overset{(2)}{2B_{\theta(2y)}}} \frac{4\pi}{\mu_0} \frac{(1\text{ fm})^2}{0.13} \right) \tag{8}$$

(1) The $B_{\theta(4Y)}$ intensity of Field #1 at the origin of Field #2 is calculated per Eq. (c) using: v_{z1}, the oscillation velocity of Field #1 calculated at z using Eq. (6); $z = 1$ fm; and h is the distance to the z-axis of Field #2, r_x.

(2) The total $B_{\theta(2Y)}$ strength of Field #2 is the sum of $B_{\theta(2Y)}$ increments for both half-fields. The sum for each half-field is calculated per Eqs. (c') and (c") using: v_{z2}, the oscillation velocity of Field #2 calculated at z using Eq. (6) plus v_{z1}; and $z = 1$ fm.

(3) F_L represents the initial Lorentz-type force between the two fields.

(4) The form of Eq. (8) demonstrates the vector summation mechanism of the B-elements joining the two fields.

Although the attractive force between two side-contacting electron fields (calculated from Equation 8 and shown in Figure 8(b)) is determined in a manner similar to the force exerted on an electron field by an applied magnetic field (Equation 7; Figure 7), the directions of the electron movements imposed by the same type of forces are opposite. In Figure 7, the direction of movement is governed by the inclination of the z-axis of the electron field. By contrast, in Figure 8(b), the z-axis is not inclined because the interaction is symmetrical; therefore, the vectors sum evenly in each electron half-field. Here the electron moves in the expected direction.

Neutron construction is an aberration of proton construction. The proton field consists of B-field vectors that rotate only clockwise, and is therefore stable. In the neutron, portions of each half-field rotate in opposite directions. Thus, when the neutron is detached from the nucleus of an atom, it destabilizes because these movements work against each other. Parts of the neutron field are ejected (creating an electron and antineutrino, popularly known as the weak interaction) to form a stable proton. Within the nucleus, the neutron field can exist in an unstable state because it is contained by the nuclear force. The neutron is created in a similar manner to the proton, except that the oscillations of the closely aligned electron fields in Figure 8(b) may be ¾ cycle (+/−) out of phase with each other rather than ½ cycle (+/−) as for the proton. Nuclear fusion is a more likely cause of neutron development that will be examined in Section 9.

5. Electron's Magnetic Dipole Field

Figure 9 shows the magnetic dipole field of an electron in its contracted state. Up to this point, this study has focused on the circumferential B_θ field occurring in the plane of each electron disk. The B_θ field is created by the spread of perpendicular fiber parts (see Figure 1) in the planes of the disks as shown in Figure 2. Owing to the Lorentz contraction of the oscillating radial fiber in each disk, the perpendicular parts are also spread along each radial fiber line in the h–z planes of the cylindrical field (see Figure 9).

This spreading in the h–z planes creates a B_z field within the cylindrical electron field in addition to the B_θ field. Because the disks are immediately adjacent to each other, the B_z field is forced outside the cylindrical field (especially during contraction of the field), forming closed loops. These loops constitute a magnetic dipole field, with B_z and B_h components. Within these loops, the exit and entry ends of the cylindrical electron field are not predetermined. The field direction can be reversed by applying a relatively strong, opposing magnetic field at one end while restraining the electron's rotation (or during admission or emission of radiation); the electron need not physically flip through 180°. The varying force, field, and v_z between two inline magnetic dipoles interacting along the z-axis are governed by Maxwell's equations 1 and 2, similar to the role of Coulomb's force in deriving the oscillating electron. In this case, the differential field changes are relative to B_h rather than B_θ and reflect differential changes in the dipole interaction force. In this case, known values of the dipole moments are used in place of Coulomb's force.

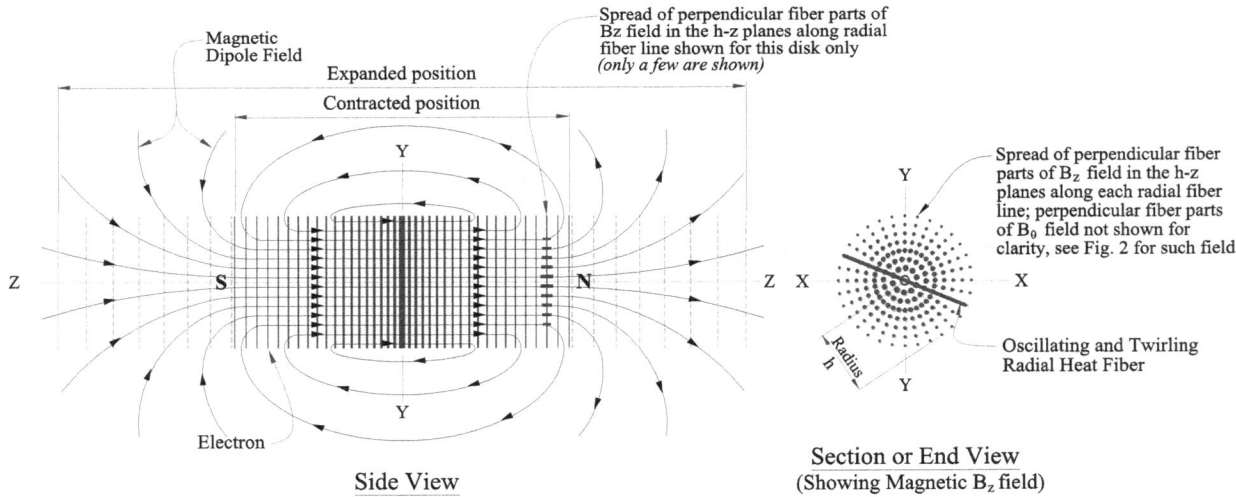

Figure 9 - The Electron and Its Magnetic Dipole Field

The dipole field lines develop alongside the B_θ field lines because both arise from Lorentz contraction of the same individual oscillating radial fibers of the cylindrical disks. Hence, the dipole field lines are synchronized with and rotate about the z-axis in the same manner and circumferential direction as the B_θ field. The mechanism of B_θ field formation is shown in Figures 2 and 3 and elaborated in Section 2. The B_θ field rotates in the angular direction of the twirling radial fibers. Because the half-circumferential B_θ fields on opposite sides of the x–y plane rotate in opposite directions, their corresponding half-dipole fields rotate oppositely.

As the magnetic dipole field rotates about the z-axis, it carries its B_h components, giving rise to a varying B_h field. As presented in this study, the magnetic interaction force between dipoles develops from the relative changes in B_h between the fields. By Maxwell's equations, this change in B_h induces a change in E_z; consequently, the fields either attract or repel. The B_h components of both fields are summed, which gives E_z, the dipole force; this mechanism is physically different from the B_h components of one field acting on an equivalent current loop of the other field via Lorentz forces. The magnetic dipole and Coulomb forces act simultaneously between (for example) two electrons. The dipole force slightly increments or decrements the Coulomb force; thus, the B_θ field derived from Coulomb's force changes slightly by including the magnetic dipole force.

6. The Electron and Electromagnetic Radiation

Figure 10 shows an electromagnetic wave forming as an electron field emits a fiber from one of its B-field disks. The mechanism is similar to that of radial fiber interactions and transits between disk spaces of an inwardly moving electron half B-field (Figure 3), except that the radial fiber (which may now be referred to as a photon) from a rear disk departs the electron field instead of advancing to the vacating forward disk space as shown in Figure 3.

The photon fibers emitted in Figure 10 have linearly motion; that is, they oscillate perpendicular to their forward motion. Circular motion occurs when a photon fiber rotates (or twirls) in planes perpendicular to its forward direction while oscillating to and fro along its axis in the same planes.

Although not evident in the *Magnetic Field Intensity of Electromagnetic Wave* in Figure 10, the magnetic field intensity appears only at current positions of the radial fibers. The continuity of the curve reflects the varying magnitude of B_x as the fiber oscillates while it traverses the z-axis. Also not shown is that, according to Equation 2(a′), the electric force potential E_y accompanying the B-field is a function of and orthogonal to the B_x component. Although E_y is not a field, Equation 2(a′) also dictates that its absolute magnitude varies in phase with the magnitude of B_x. The electromagnetic wave is solely constituted by the B-field component, which drives the orthogonal E-potential.

An electron field emits radiation when it transitions from a higher to a lower energy level in an atom or molecule and also when it accelerates. The higher the energy level (less binding energy) of the electron in an atom, the further is its position from the nucleus. Interconnected with a proton field *(examined in Section 7 and shown in Figure 11)*, an electron field oscillates in tandem with the proton. Although the energy of the electron at a particular level is constant, its potential and kinetic energy components continuously vary as the kinetic energy is varied by the coupled electron/proton oscillations.

As the electron descends to a lower energy level in the atom or molecule, it becomes more tightly bound to a proton(s) in the nucleus; that is, its potential energy decreases and it becomes captured more tightly by the Coulomb force. Figure 11 shows the electron/proton interaction. During this interaction, the electron field is drawn into the proton field, compressing the right half-field of the electron field. With less space available, the electron is forced to shed one or more of its radial fibers. The greater the energy difference between the previous and current state of the electron, the greater the field compression, and the higher the energy and quantity of emitted radiation. The lost energy (emitted as radiation) equals the frequency of the emitted radiation ν times the Planck constant h. These energy changes are intermittent rather than continuous. Below, we relate the non-orbiting model in this study (Figure 11 in Section 7) to the Bohr model of the hydrogen atom.

In Bohr's model of the hydrogen atom (an electron orbiting a proton), electron energy is lost (as radiation) in permitted states or orbits as the electron approaches the proton from one discrete principal state to another. The principle energy levels are denoted by natural numbers n, where $n = 1$ represents the ground state of the electron, where radius r equals the Bohr radius a_0. Other allowed orbital radii equal $n^2 \times a_0$. As mentioned above, the energy emitted in a transition from a higher to a lower permitted state is given by vh.

For an electron not transitioning between states, Bohr's model assumes that 1) the orbiting electron has a centripetal acceleration induced by Coulombic attraction between the electron and the central protons, and 2) the classical orbital angular momentum of the electron is quantized by $n \times h/2\pi$. Combining these two criteria and assuming a static electron (rather than the dynamic one implied by item 1) subjected to an outward force equivalent and opposite to that responsible for its centripetal acceleration, we obtain the following force-balance equation in the radial direction:

$$(\text{Outward Force} = n^2h^2/4\pi^2 m_e r^3) = (\text{Inward Force} = e^2/4\pi\varepsilon_0 r^2) \tag{9}$$

This equation gives the permitted radii of the electron orbits for each principal quantum number "n." The Bohr radius a_0 is the radius for $n = 1$. Because n is independent of r, the forces balance only in the principle (n) states. Between such states, the outward and inward forces are proportional as the electron moves radially. Both forces are external to the electron. The work associated with each force as the electron moves in the radial direction from infinity to a permitted radius is the integral of the incremental work performed by each force between these limits. The algebraic sum of the work performed by both forces gives the total energy of the electron at such radius. Because the outward force is proportional to but less than the inward force between states, the absolute work performed by the inward force is twice that performed

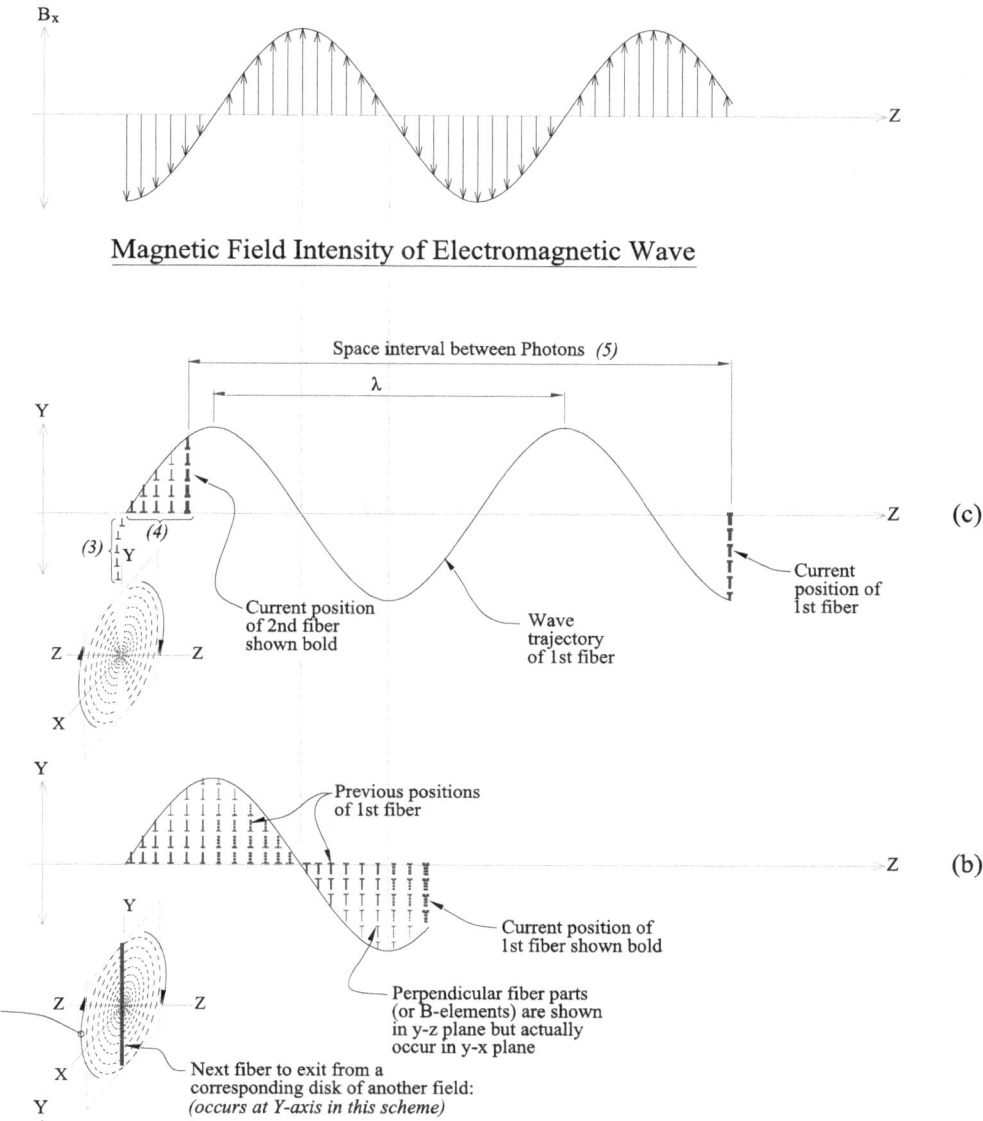

Magnetic Field Intensity of Electromagnetic Wave

Fiber Transitions from Electron Disk to Electromagnetic Wave in sequence (a) to (c)

Figure 10 - The Electron and the Electromagnetic Wave

by the outward force. Thus, the energy lost by the electron to radiation during transition equals the stored energy *vh* (i.e., the work performed by the outward force).

Because the predictions of Bohr's model are consistent with experiment, the quasi-outward force concept in Bohr's model is validated. However, in this study (as modeled in Figure 11), the outward force is substituted by an internal force, with which the electron field resists the external inward Coulomb force during contraction and compression as it advances toward the proton. The work performed by the internal outward force equals the kinetic energy of the radiation emitted at the permitted quantum levels of the electron in the hydrogen atom. During contraction of the field, the internal force is mobilized by space requirements and/or restrictions between electron disks. Radial fibers from rear disks attempting to advance and interact with fibers of forward disks may be resisted by lack of space between disks, large energy differences, excessively rapid interaction rate or non-synchronized twirling, and may instead be discharged from the electron field as radiation, as depicted in Figure 10. For example, assume that disk 3 in Figure 3 was admitted into the field during its outward oscillation phase. Then, as the field contracts, one of the abovementioned factors prevents disk 3 from properly interacting with disk 4, and the newly admitted fiber generating disk 3 is discharged as radiation instead of being retained.

Now assume that the fiber of disk 3 is discharged as radiation when the electron field in a hydrogen atom moves inward from a higher principal quantum level to its natural principal quantum level. During this transition, the internal work performed in the electron field by its internal outward force, owing to one of the abovementioned factors, is the energy of the radiation emitted from disk 3, and it is associated with the failed interaction between disks 4 and 3. The internal work performed in the transition equals the kinetic energy *vh* of the emitted fiber (or photon).

The relationship between varying fiber energy and B-disk density along the z-axis (see Section 2) is equally applicable to radiation: *As the B-disks of each half-field retract toward the x–y plane, the effective varying diameter of each half-cylindrical field gradually decreases to a minimum at the x–y plane. This occurs as follows: The increasing concentration of B-elements accumulating in the inner zone draws the outer disks together and concentrates them near the inner zone. The high concentration of perpendicular fiber components (B-elements) in the inner zone (particularly near the z-axis) attracts fibers in rear disks, whose oscillation ranges accordingly decrease. Thus, the diameter of each half-field is minimized at the x–y plane. The vector summation of this process is given by Equation 1(b). Because all fibers continue to oscillate at near-light velocities during transit of the half-fields, their oscillation frequencies (and thus their energies) increase with their decreasing oscillation ranges during the approach to the x–y plane.* Thus, the kinetic energy of individual discharged fibers is increased if their corresponding disks lie closer to the electron center (in the above example, the fibers are discharged from disk 3). Because the oscillation range is less for in-close disks, these fibers are ejected with higher frequency (equivalently, with smaller wavelength). Hence, as electrons in hydrogen atoms transit to increasingly lower energy levels (lower n but greater binding energies), the oscillation range of the disks ejecting the fibers become progressively smaller, in turn supplying the emitted fibers with higher frequencies and energies. Equation 10 equates the work of the internal outward force to the energy of the emitted radiation in hydrogen atoms. Note that the radiation frequency v inversely varies with n^2.

$$h^2/(8\pi^2 m_e n^2 a_0^2) = vh \tag{10}$$

The mechanism by which an electron field admits radiation is the reverse of radiation emission. Incoming radiation from sources such as heat, light, or electrical current excites the electron field by addition of radial fibers (photons). This expands the field and causes the electron center to move outward from the atom's nucleus. The admitted radial fibers oscillate at a frequency that allows their insertion between two disks. Fibers oscillating at incompatible frequencies pass through or are deflected. The fibers entering between disks form new disks, and impart their kinetic energy to the external energy of the electron relative to the nucleus. This mechanism also explains the photoelectric effect, where the electron field detaches from the nucleus and departs with the balance of the energy of the admitted photon.

Once the electron field has admitted radiation to occupy a higher energy level in the atom, it may reject further incoming radiation if the fiber frequencies are now incompatible. In some cases, electron fields subjected to a constant radiation source establish equilibrium between radiation admission and emission, whereby they oscillate outward and inward, respectively, along their z-axis to different energy levels relative to the atom's nucleus.

The frequency of radial fibers (radiation) emitted from an electron half-field (Figure 2) depends on the location on the z-axis from which they are emitted. This phenomenon can be roughly understood as follows. By Equation (c), B_θ varies inversely with z^3 (at relatively small h), which primarily influences the oscillation range (disk diameter D) of the fibers, while B_θ varies inversely with disk area D^2. Thus, z^3 is roughly proportional to D^2. The oscillation range of a fiber, and thus its disk diameter D can be determined from its frequency v by the following relationship: $D = c/2v$. The most energetic electron fibers, with the shortest oscillation stroke, are x-rays; thus, x-rays are emitted closest to the origin of the electron. If a median-energy x-ray (3.75×10^{17} Hz) is presumed to occur at z = 1 fm, then the fibers of lower frequencies can be roughly organized on the z-axis according to the proportionality relationship $z^3 = 6.25 \times 10^{-27}$ D^2. Thus, the oscillation range D of a typical light fiber (6×10^{14} Hz) is 2.5×10^{-7} m and is located on the electron's z-axis at $z = 7.5 \times 10^{-14}$ m.

7. Hydrogen

Figure 11 illustrates an electron field interacting with a proton field in a hydrogen atom, from an arbitrary initial position (a) to the ground position (b). Under attractive Coulomb forces, the separation distance r between the fields decreases to its minimum at a_0 (+/-), denoting the ground state, as outlined beneath. In reverting to the ground state, the excited electron at position (a) radiates some of its fibers during the interaction. The energy of the state that an electron field moves to is a mean energy level since the electron oscillates about this level, in tandem with the proton field, along the z-axis; this is examined below.

The longitudinal axis of each field aligns along the z-axis. When both inner half-fields approach or retract along the z-axis (relative to their respective centers), the electron and proton obey the left-hand and right-hand rules, respectively, and their B-fields align in the same direction. The mechanism that causes the cylindrical field of an individual electron to expand and contract—namely the vector summation of the B-field disk components (Equation 1a)—causes the proton and electron fields to unconditionally attract via Coulomb's force because their B-fields are in the same direction. In this case, the B-field vectors of the overlapped disks and their corresponding radial fibers accumulate and draw the centers of both fields toward each other. This mechanism acts during outward oscillatory movement of both fields when their inner and outer half-fields become contracted and expanded, respectively (Figure 11(b)).

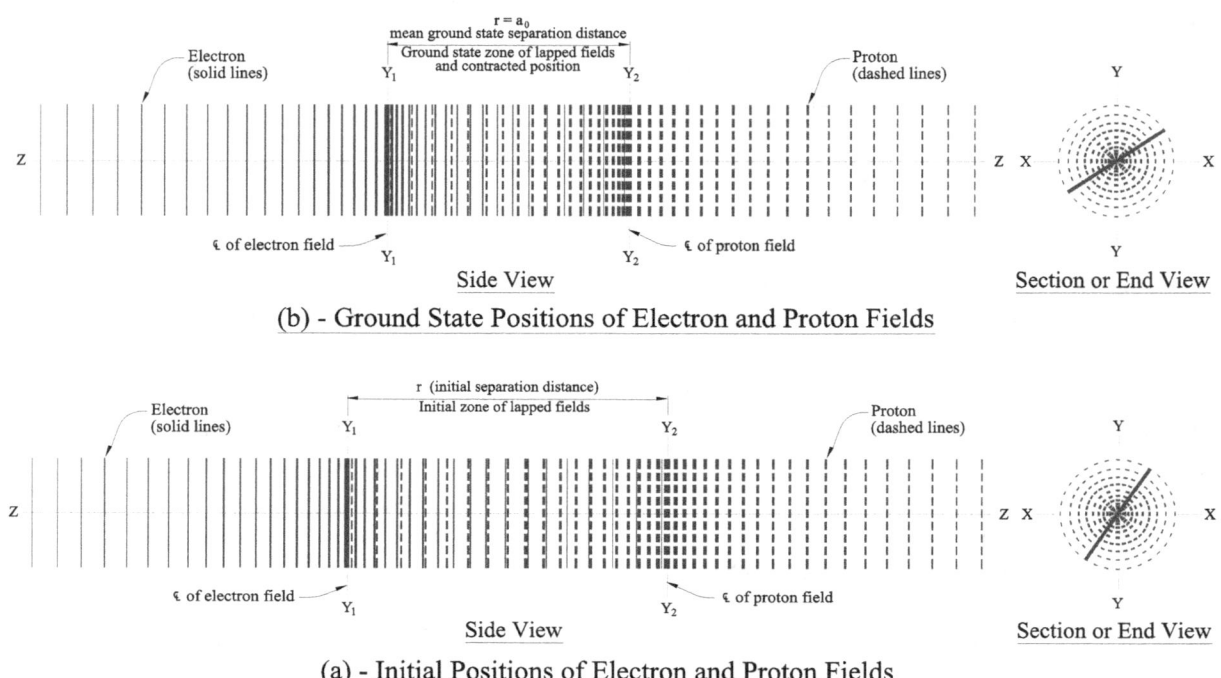

(b) - Ground State Positions of Electron and Proton Fields

(a) - Initial Positions of Electron and Proton Fields

Figure 11 - Interaction of Inline Electron and Proton Fields
Hydrogen Atom

Similarly to individual electron fields, during the above interaction, the outermost disk of each inner half-field crosses the x–y plane of the other field. Within this zone, the direction of the B-vectors of the overextended part of each inner half-field opposes the B-vector direction of the outer half-field of the other field, respectively, both of which are undergoing an outward oscillatory movement at this instant. Consequently, both inner half-fields cease their outward movements, reverse, and then move toward their respective x–y planes. Then, as the innermost disks of each inner and outer half-field of both electron and proton cross their own x–y plane, their directions reverse and the bicycle repeats. The outer half-fields of the electron and proton generally remain uncompressed but oscillate in unison and in the opposite direction with their respective inner half-fields along the z-axis.

During this interaction, the mechanism propelling the outward oscillatory movement of both inner half-fields during their simultaneous contraction is that depicted in Figure 3. In this case, however, the disks do not move toward the centers of their fields, but instead, the center of each field moves toward the outer transiting disks of its inner half-field. Once the outermost disk of each inner half-field begins to cross the x–y plane of the other field, the oscillatory movement reverses as described above.

As the inner halves of the electron and proton fields merge under the summation mechanism of their B-vectors, the individual disks of both half-fields must intercept as both fields simultaneously contract. During this movement, the radial fibers responsible for these B-vectors must either pass through or interchange with each other within the interference zone. For example, a fiber forming a part of a proton disk may detach during the interception and join the electron disk, while the electron disk correspondingly loses a fiber to the proton disk. Fibers not interfering process unimpeded. The varying B-field and speed of each field at the instant of their initial positions (Figure 11(a)) can be calculated from Equation 6 and Equation (c), from which the oscillation velocities and magnetic field strength "B" are derived for individual oscillating electrons.

A. A Hydrogen molecule H₂ constitutes a pair of bound hydrogen atoms, realized when the half B-fields of electron 2 and proton 1 overlap (where such fields are indicated in Figure 12). The bonding process is governed by the same vector summation mechanism that binds the electron and proton in the hydrogen atom. Figure 12 shows the ground state of the molecule; the three inner sections, where the half-fields overlap, are contracted to separation distances of $r = a_0(+/-)$, while the outer half-fields of electron 1 and proton 2 remain relatively expanded.

The fields of both electrons and both protons oscillate in unison along the z-axis, as described above for the individual hydrogen atom. Hence, as occurs in the atom, each proton and electron half-field (relative to their x–y planes) simultaneously moves outward (or inward). During the outward movements, the centers of all four fields are drawn toward each other as the three inner sections contract. The contraction is symmetrical about the midpoint of the zone of overlap between the half-fields of proton 1 and electron 2. The two outermost half-fields also expand outward, but their net movement is inward due to the contraction of the three inner sections. During an inward oscillatory movement of each half-field (relative to its x–y plane), the movements are reversed, and the molecule expands similarly to the atom.

No more than two hydrogen atoms can connect end-to-end along the z-axis, as shown in Figure 12. For example, the electron half-field of a third atom may attempt a linear bond with the outer half-field of proton 2; however, the bonding would fail for the following reasons. As explained above, if the molecule contracts along the z-axis in Figure 12, the outer half-field of proton 2 must translate inward. For bonding, the half-field of the third electron must translate in the same inward direction. However, governed by the right-hand and left-hand rules respectively, the B-fields of the translating proton and electron rotate in opposite directions,

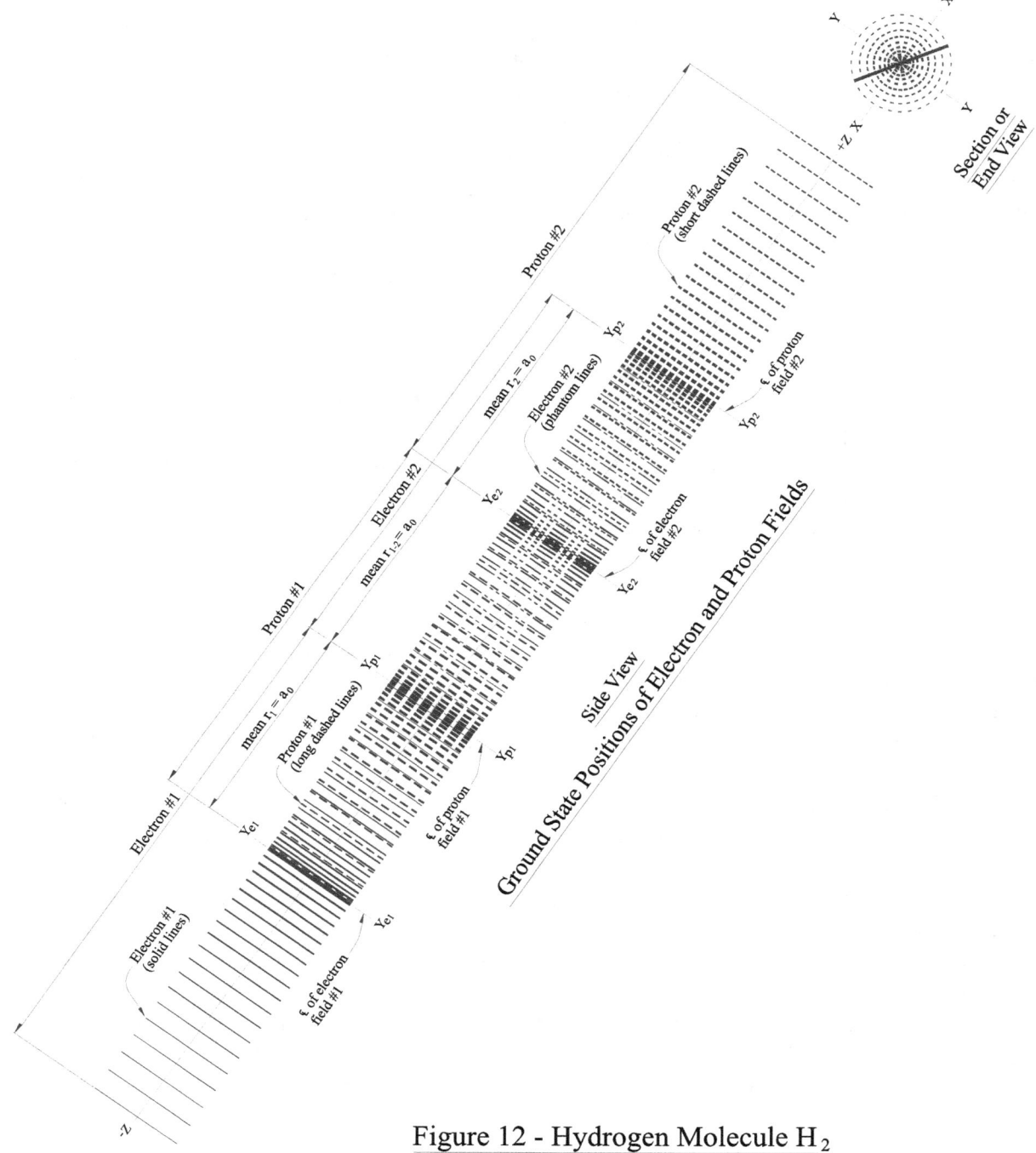

Figure 12 - Hydrogen Molecule H$_2$

thereby rejecting the third atom. In effect, this mechanism prevents two atoms from sharing two electrons in the same quantum state. This phenomenon is known as the Pauli Exclusion Principle. The position, orientation, and circumferential direction of the electron's B- field describe its quantum state. Hence, all of the B-fields presented in this study are fermions.

B. Hydrogen gas structure is schematically illustrated in Figures 13(a) and (b) below. The gas structure is composed of H_2 molecules interconnected in various orientations. The outer half-field of an electron of one molecule makes a skewed connection to the outer half-field of a proton of another molecule; and the outer half-field of the proton of the first molecule makes a skewed connection to the outer half-field of an electron of a third molecule. Multiple repeats of these connections form the pattern shown in Figure 13, in which four skewed molecules are connected to each end of each molecule oriented parallel to the z-axis (referred to as inline molecule I). Of the four skewed planar molecules, the two molecules lying parallel to the x–z plane and to the y–z plane are referred to as skewed molecules S and connector molecules C, respectively. All four molecules are oriented at approximately the same angle, referenced from a line through the intersection of the connection and parallel to the z-axis.

The four skewed H_2 molecules can connect endwise to a linear I molecule as follows: Unlike the preceding case forbidding the linear connection of a third atom, the movement of a skewed molecule need not be synchronized with the small translational movement of the oscillating linear molecule. Thus, each skewed molecule can oscillate in unison but in opposite directions, with the linear molecule in the absence of common translation. Connection is maintained only by slight rotation. Hence, skewed connections are allowed because the outer halves of the B-fields of the skewed and linear molecules never oppose. The skewed connection is much weaker than the linear connection between the two atoms of the H_2 molecule because the outer half-fields of the skewed and linear molecules are only partially engaged and contact at an angle. The opposing B-fields of other skewed molecules connected to the same end of the linear molecule also weaken the connection.

The orientations and gaseous structure of H_2 molecules shown in Figures 13 (a) and (b) are based on the hydrogen spectrum and the energy levels of the 3s to 2p, 2p to 1s, and 2s to 2p transitions. When a magnetic field of sufficient strength is applied parallel to the z-axis, these energy levels are split by the Zeeman Effect. Figure 13(c) is a schematic of the energy levels and transitions between these states in the absence and presence of an applied magnetic field.

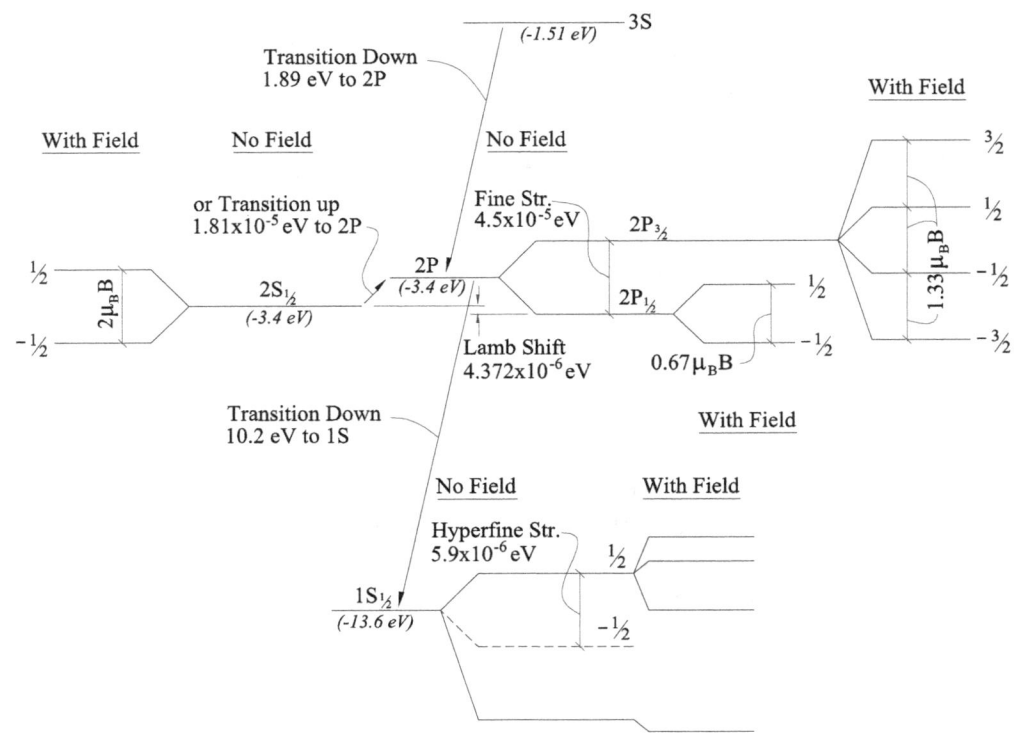

(c) - Energy Levels and Transitions

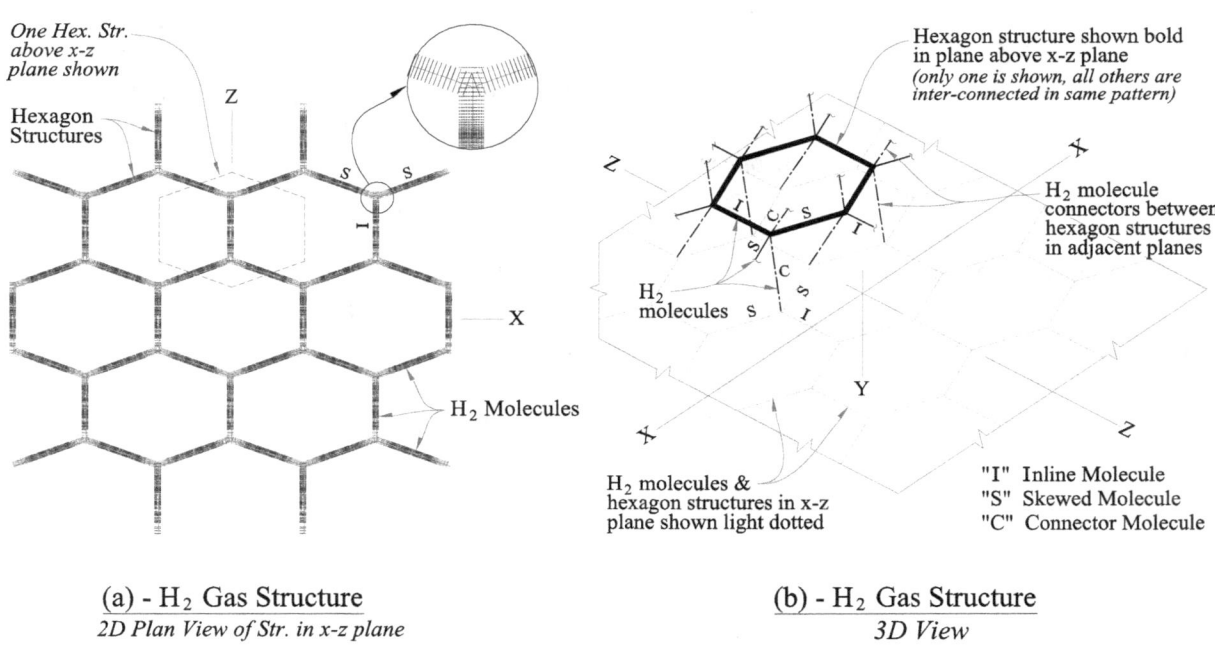

(a) - H₂ Gas Structure
2D Plan View of Str. in x-z plane

(b) - H₂ Gas Structure
3D View

Figure 13 - Hydrogen Gas

The gas structure is composed of interconnected hexagons bounded by inline I molecules parallel to the z-axis, two skewed S molecules parallel to the x–z plane, and two skewed connector C molecules parallel to the y–z plane. The S and C molecules are connected to each end of the I molecules. The planar hexagonal lattices lie parallel to the x–z plane, and are stacked by the C molecules in the y–z plane.

The unit hexagon immediately above the hexagonal lattice in the x–z plane shown in Figure 13(b) is offset from its lower counterpart in the x-direction by half a unit hexagon. All other hexagons (not shown) in the upper plane are offset in the same way. In the next plane above (two planes above the x–z plane), the hexagonal lattice again aligns with the lattice in the x–z plane. Hence, the hexagonal lattices in planes parallel to the x–z plane align with the corresponding lattices in alternating planes. The gas structure and behavior is independent of whether the S or C molecule is oriented parallel to the z-axis because four skewed molecules are connected to each end of a parallel molecule at similar orientations.

C. Fine Structure Examination: As discussed above for the hydrogen atom, the overall length of the H_2 molecule (see Figure 12) increases and decreases when it admits and releases radiation, respectively, into and from its three inner sections. This phenomenon occurs in all molecules undergoing transitions between principal states although the lengths of the three inner sections cannot decrease below the ground state length. In the molecular structure of Figure 13 (a) and (b), only those molecules offering less resistance will admit tiny quantities of radiation (such as occurs in fine structure splitting; see Figure 13(c)); more resistant molecules will exclude these tiny quanta. Less resistant molecules can expand their length with relatively little or no restraint by adjacent molecules; resistant molecules are much more restrained by adjacent molecules. If the restricted molecules expand, the restraining adjacent molecules are required to expand also.

The I molecules offer the least resistance to tiny radiation quanta because their lengths can increase without affecting the lengths of the four molecules attached at each of its ends. The enlargement in Figure 13(a) shows the bonding details of two of the four attached molecules. The I molecule can incrementally expand in the z-direction unhindered by the S and C molecules attached at its ends. The C molecules can also admit such radiation because the only restriction on their expansion is that the planar hexagonal lattices (parallel to the x–z plane) move apart in unison. As the lengths of the C molecules increase, they rotate in planes parallel to the y–z plane and adjust their contact angles with the I molecules. Because small-length expansions of the S molecules require rotation of their ends into the ends of the C molecules, rotation of the S molecules is subject to a Lorentz-type restraint (see enlargement in Figure 13(a) and Figure 13(b)). Consequently, the S molecules, unlike the I and C molecules, will reject small quantities of radiation.

During a transition from the 3s state to 2p state, the molecules emit radiation and undergo length contraction, with slight changes in the orientations of the C and S molecules relative to the I molecules. Although the energy of the radiation emitted by all molecules is the energy difference between these principal states, the I and C molecules retain the small energy differences induced by fine structure splitting, while the S molecules do not (see Figure 13(c)).

During a 3s to 2p transition, the I and C molecules move to the $2p_{3/2}$ state. The I molecules then drop further to the $2s_{1/2}$ state, while the S molecules move to the $2p_{1/2}$ state. The lower energy of the S molecules in the $2p_{1/2}$ state relative to the I molecules in the $2s_{1/2}$ state is likely attributable to the restraint on the S molecules examined above.

In the hydrogen gas structure of Figure 13, each I molecule is associated with two C molecules and two S molecules. During a 3s to 2p transition, roughly 40% of the molecules move to the $2p_{1/2}$ state and 60% move to the $2p_{3/2}$ state. Of the latter, 33% then move to the $2s_{1/2}$ state. The I molecules in the $2s_{1/2}$ state, the C molecules in the $2p_{3/2}$ state, and the S molecules in the $2p_{1/2}$ state participate in the Zeeman Effect when the gas is exposed to a magnetic field.

The fine structure splitting of energy levels in the presence and absence of the magnetic field, explainable in terms of small energy differences between the S molecules and remaining molecules (I and C) in the 2p state, is associated with the interaction energy of the intra-magnetic dipole fields of the H_2 molecules. As explained above, the I and C molecules can extend relatively freely, enabling them to admit small additional radiation quantities. In turn, these molecules acquire greater energy via stronger intra-magnetic dipole field strengths (which elevates the electrons to the higher energy state $2p_{3/2}$) than is possible without additional radiation absorption. The strength of the electron dipole fields of molecules in the $2p_{3/2}$ state are double those of molecules in the $2p_{1/2}$ and $2s_{1/2}$ states. Currently, the small energy increase associated with fine structure splitting is modeled and measured as the dipole field or magnetic moment of the electron (owing to its quantized spin angular momentum) under an internal magnetic field created by the quantized orbital angular momentum L of the electron; if the two fields are anti-parallel then the electron exists in the higher energy state $2p_{3/2}$.

Both electrons and both protons in the H_2 molecule of Figure 12 possess a magnetic dipole field similar to that presented in Figure 9, although the dipole fields of the electrons are much greater than those of the protons (as measured by their magnetic moment constants or multiples of these). The construction of a dipole field and the interaction between two dipole fields has been described in Section 5: *Electron's Magnetic Dipole Field*. For the H_2 molecule in Figure 12, the dipole field of proton 1 interacts with the dipole field of both electrons 1 and 2. However, because the proton resides between the two electrons, this interaction generates no combined net change in the electron energy levels, although it impacts the hyperfine splitting of the energy level of the electron in the $1s_{1/2}$ state, as examined below. In particular, the dipole field interaction between electron 2 and proton 2 increases or decreases the energy level of the electron depending on whether the fields align anti-parallel or parallel, respectively (this also affects the hyperfine structure splitting of the energy level for the electron in the $1s_{1/2}$ state, this is examined below). The strongest magnetic dipole field interaction, which impacts the splitting of the energy levels shown in Figure 13(c), occurs between the two electron fields.

As presented above, admission of small additional radiation into the electrons of the H_2 molecule at the 2p level can raise their energy to the $2p_{3/2}$ level by doubling the intrinsic dipole field strength of each electron. In the $2p_{3/2}$ state, the dipole fields of the two electrons are anti-parallel. The fine structure splitting energy (4.5×10^{-5} eV) depicted in Figure 13(c) can be determined by calculating the work required to reverse (or equivalently flip) the direction of the magnetic dipole field of one electron under the influence of the magnetic dipole field of the other electron (where the electron centers are separated by $8a_0$ because both are in the $n = 2$ principal state) and adding the Lamb Shift to the result. During this interaction, the electrons emit the small additional radiation, and the dipole fields reorient to a parallel alignment, while their strengths return to their intrinsic levels in the $2p_{1/2}$ state.

The fine splitting of the $1s_{1/2}$ energy levels in Figure 13(c) results from combined interactions between the four intrinsic magnetic dipole fields within the H_2 molecule of Figure 12. The dipole field of proton 1 interacts with the dipole fields of electrons 1 and 2, while that of proton 2 interacts with the dipole field of electron 2. These interactions are respectively described in the *first* and *second case* below.

Considering the first case: Suppose that the dipole fields of the two electrons are anti-parallel with electrons 1 and 2 oriented in the $-z$ and $+z$ directions, respectively. The dipole fields of proton 1 and electron 2 are both oriented in the $+z$ direction. In this case, the r_1 separation increases and radiation is admitted into this zone, while the r_{1-2} separation decreases with radiation emission. The energy added by the absorbed radiation in zone r_1 contributes to the upper energy level of $1s_{1/2}$ in Figure 13(c), while the lower energy level is solely contributed by the energy lost as radiation in zone r_{1-2}. The average energy level of both zones is indicated by the dashed line in the figure. This line also represents the energy of interaction between parallel dipole fields of an electron and proton in a single hydrogen atom, while the upper level corresponds to the anti-parallel equivalent.

Considering the second case: Now consider that the dipole field of proton 2 is oriented in the $-z$ direction, while that of electron 2 orients in the $+z$ direction as before. In this case, the two fields are anti-parallel; the r_2 separation increases and radiation is admitted into the zone. The energy added by the radiation in zone r_2 contributes to the upper energy level of $1s_{1/2}$ in Figure 13(c) and is additional to the contribution in the first case. The hyperfine energy splitting (5.9×10^{-6} eV) can be determined by calculating the work required to reverse (or flip) the direction of the dipole field of proton 2 (where electron 2 and proton 2 are separated by a_0 because the electron exists in the $n = 1$ principal state). As the proton dipole field reverses its direction, the electron emits radiation at the hyperfine splitting energy.

The energies split by the interactions in the above two cases can be further split in a special case of the Zeeman Effect, as shown for the $1s_{1/2}$ level in Figure 13(c). Although the magnetic field dipole interactions in the H_2 molecule are more energetic between the two electrons than between an electron and a proton, the energies remain much smaller than and must be included in the principal state energy at $n = 1$; otherwise, this energy level would undergo observable splitting. This implies that the dipole fields of both electrons are consistently oriented in their same directions in the ground state of all hydrogen gas molecules.

8. Gravity

A. Microscopic Field: The overlap of certain B_θ fields of neighboring H_2 molecules gives rise to a gravitational field within the H_2 gas structure of Figure 13(b). In each molecule, the participating B_θ field is perpendicular to the z-axis, as configured in Figure 12. The vector summation of these certain B_θ fields between neighboring molecules leads to a derivative of the Lorentz-type attractive force (Equation 2a′)—the force of gravity—between and beyond the neighboring gas molecules.

Figure 14 shows a cross section through two neighboring H_2 molecules in the hydrogen gas, both oriented with their +z-axes perpendicular to and projecting into the diagram. The cross section may represent two I molecules or the vertical components of two C molecules *(the horizontal components of the C molecules are shared by other sections, however, the net effect of combining the two sets of components gives a resultant modeled along their z-axes)*. The section is taken through the electron B_θ field of the left molecule and the proton B_θ field of the right molecule. As previously examined, the B_θ field of electrons and protons arises either from oscillation of their half-fields to and fro along their z-axes or by translation of their half-fields along their z-axes. The attractive Lorentz-type force in the x-direction, resulting from vector summation[5] between the electron and proton fields, can be calculated by Equation 2(a') and by replacing the applied perpendicular B term with the B_y intensity *(the sum of B_y increments from theoretical infinity to the point of measurement)* of one field (say the electron field) at the origin of the other field (the proton field) as follows.

$$F_x = -v_{zp}\,|e_p|\,B_{ye} \qquad (11)$$

The v_{zp} term is the velocity along the z-axes of the proton field (relative to the electron field) at its origin ($z = \pm 1$ fm when considering oscillation of the half-fields or $z = 0$ when considering translation of the whole field). B_{ye} is calculated per Equation (c) as follows. Use $z = \pm 1$ fm when considering oscillation of the half-fields or $z = 0$ when considering translation of the whole field, $h = r_x$, and v_{ze} = velocity of the electron field along its z-axis and at its origin ($z = \pm 1$ fm when considering oscillation of the half-fields or $z = 0$ when considering translation of the whole field). The half B_θ fields of either an electron or proton oscillate in unison but in opposite circumferential and translational directions and with equal speed along the z-axis. Thus, outside the locality of the interaction shown in Figure 14, the summation of the B_θ increments and therefore B_y increments of either the electron or the proton field is zero since their half-fields oppose and cancel each other. As such, no gravitational field would occur from the electron and proton interaction shown in Figure 14 due to the oscillations of their half B_θ fields to and fro along their z-axes.[12]

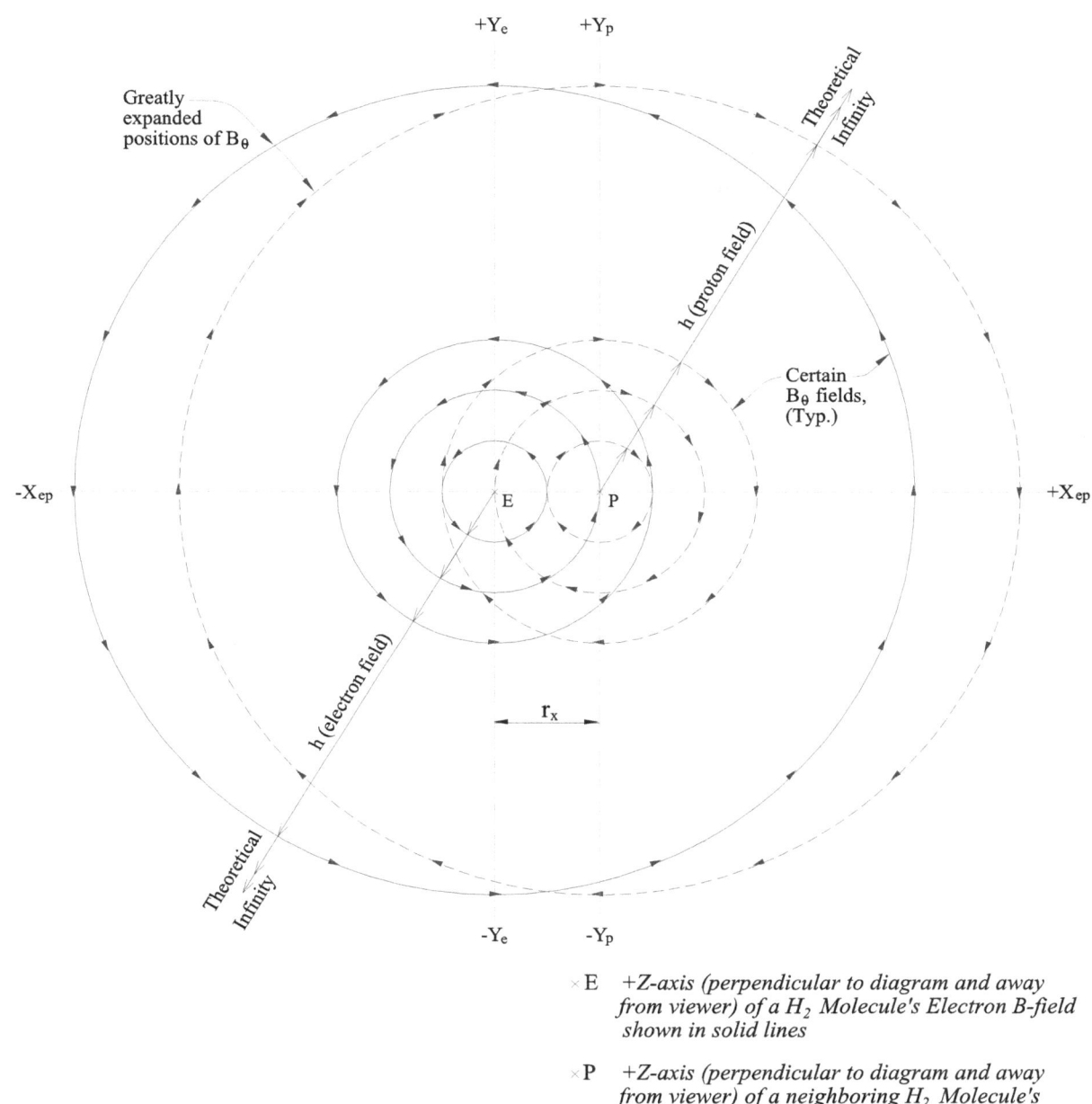

+Y_e   +Y_p

Greatly
expanded
positions of B_θ

Theoretical
Infinity

h (proton field)

Certain
B_θ fields,
(Typ.)

-X_ep   +X_ep

E   P

h (electron field)

r_x

Theoretical
Infinity

-Y_e   -Y_p

×E   *+Z-axis (perpendicular to diagram and away
from viewer) of a H₂ Molecule's Electron B-field
shown in solid lines*

×P   *+Z-axis (perpendicular to diagram and away
from viewer) of a neighboring H₂ Molecule's
Proton B-field shown in dashed lines*

Figure 14 - Hydrogen Gas Created Gravity Field
H₂ Molecules' Electron/Proton Interaction

Hence, the participating B_θ field mentioned in the opening of this section and depicted in Figure 14 likely arises from small oscillatory translations at the origins of the electron and proton fields. As previously explained, as each oscillating electron (or proton) half-field moves inward toward the origin (the x–y plane; see Figure 2), a portion protrudes into the opposite half-field. The field movements in the crossover zone *(z = −1 fm to +1 fm)* are schematically illustrated in Figure 4. Portions of the half-fields extend into each other's space until the repulsive force acting on their innermost disks (owing to the opposing direction of their B_θ fields) reverses their direction along the z-axis.[1] However, the innermost disks of each half-field move across the x–y plane at slightly different times. Thus, one disk is instantaneously restrained and infinitesimally delayed at the crossover by the opposing disk. This delay causes an infinitesimal increase in the velocity (v_z) of the delayed disk at the crossover, relative to the opposing disk. As the oscillatory movement of the half-fields reverses outward (and the overlapped portions of the half-fields return to their home sides of the x–y plane), this process repeats at the crossover zone, but with the opposite hand. The delayed innermost disk moves in unison with the opposing disk and instantaneously crosses afterward at infinitesimally greater velocity (along the z-axis). This tiny velocity difference (v_z) during the field's crossover causes oscillatory translation ($\pm v_z$) of the whole field.

The above-mentioned velocity difference (v_z) is proportional to the densities of the disks/fibers in the crossover zone. Because the time delay is greater at higher disk/fiber density, the delayed disk requires a higher velocity to compensate for the lost time, if both innermost disks are to maintain equal and opposite displacements into the opposite sides of the x–y plane. The increased time delay is caused by the greater density (or energy) of the intervening fibers. The delayed innermost disk receives an extra "push" from the compression of the other disks in the crossover zone on the same side of the x–y plane. As the density of fibers in the crossover zone increases, the oscillation strokes of the fibers shorten as described in Section 2, and each fiber acquires more energy. As stated above, this lengthens the crossover time of the first innermost disk, imparting a greater v_z to the second disk.

During inward oscillatory movement of the two half-fields, the field infinitesimally translates in one direction along the z-axis at a net velocity equaling the difference in oscillation velocities between the two half-fields in the crossover zone. As the inward movement reverses to outward in this zone, the field infinitesimally translates in the opposite direction with equal and opposite net velocity. Thus, not only do the electron (or proton) half-fields oscillate in opposite directions along the z-axis, the whole field undergoes tiny translations along the z-axis in unison with the oscillating half-fields.

Each infinitesimal translation of the electron (or proton) is associated with a small B_θ field. Because the translational movements are oscillatory, the circumferential directions of the associated B_θ (hereafter-denoted $B_{\theta G}$) field also oscillate. The $B_{\theta G}$ field, a derivative of the electron's B_θ half-fields, arises from the tiny translational movements of both half-fields; thus, the distribution of $B_{\theta G}$ field intensity derives solely from these movements *(effectively representing the net movements of the whole field because the oscillatory movements of the half-fields cancel)*. Summing the y-tangential components of each incremental circumference of each disk of a $B_{\theta G}$ electron field with those of a neighboring $B_{\theta G}$ proton field (as shown in Figure 14) yields an attractive Lorentz-type force F_X between the fields. See Equation 11 for the force acting on a translating field calculated using the parameters specified for translation. This force, referred to as F_{XG}, denotes the gravitational force between the fields. Equation 11 can be rewritten as follows:

(12a) $F_{XG} = v_{zp} \, |e_p| \, B_{YGe}$; and from Eq. (c),   (12b) $B_{YGe} = \dfrac{v_{ze}}{c^2} \, \dfrac{|e_e|}{4\pi\varepsilon_0} \, \dfrac{1}{(r_x)^2}$ (12)

$$\text{Thus:} \qquad F_{XG} = \overbrace{\left[\dfrac{v_{zp} \, v_{ze}}{c^2} \underbrace{\left(\dfrac{|e_p \, e_e|}{4\pi\varepsilon_0} \, \dfrac{1}{(r_x)^2} \right)}_{\text{Coulomb's } F_C} \right]}^{\text{Lorentz-type force } F_L}$$ (13)

Because the interactions are symmetric, Equations 12(a) and 13 are positively valued. The ratio of the Coulomb to gravitational force between an electron and proton field separated by r_x is approximately 2.27×10^{39}. For F_{XG} to equal $(1/(2.27 \times 10^{39}) \times F_C)$, the term $(v_{zp} \, v_{ze})/c^2$ in Equation 13 must equal the same ratio: $1/(2.27 \times 10^{39})$. As stated above, v_{zp} and v_{ze} are the differential net translation velocities of the proton and the electron fields, respectively, and are proportional to the quantities commonly known as the "masses" of the proton and electron, respectively (these relationships are examined below). Setting the velocity ratio equal to the mass ratio gives $v_{zp} = 1836 v_{ze}$. Then $1836 v_{ze}^2/c^2$ must equal $1/(2.27 \times 10^{39})$, giving v_{ze} approximately 1.5×10^2 fm/s, from which v_{zp} equals $\sim 2.7 \times 10^5$ fm/s. Because v_{zp} is much greater than v_{ze}, it can also be viewed as the velocity relative to the electron field, consistent with Equation 12(a).

Equation 13 can be expressed more intuitively as a derivation of Equation 12(a) as follows: $F_{XG} = B_{YGe} \, B_{YGp} \, (4\pi/\mu_0) \, (1 \, fm)^2$ where B_{YGe} is calculated from Equation 12(b) and the term $[B_{YGp} \, (4\pi/\mu_0) \, (1 \text{ fm})^2]$ equals $v_{zp} \, |e_p|$ in Equation 12(a). The latter term is obtained from Equation (c), substituting the quantities v_z and e with v_{zp} and e_p, respectively, and setting $z = 0$ and $h = 1$ fm. Because B_{YGp} is expressed in terms of $h = 1$ fm, the required area unit is $(1 \text{ fm})^2$, or 1×10^{-30} m^2; thus, F_{XG} is independent of choice of h. The product of the B strengths in the above equation reflects the vector summation of the y-tangential components of the two $B_{\theta G}$ fields responsible for the attractive force between them.

The gravitational force between the fields is measured as the attraction between the B_{YG} components of the electron and proton $B_{\theta G}$ fields. From this, their individual masses (which are associated with their gravitational weights) are determined. Without the differential net translational velocity ($\pm v_z$) of both fields, the $B_{\theta G}$ fields would not exist, and thus there would not be a measurement commonly known as a mass. Because the differential net velocity of both fields is proportional to the gravitational force between them (see Equation 13), it must also be proportional to their mass measurements.

The measured inertial masses of the electron and proton fields equate to their measured gravitational masses. The inertial mass of a field defines its resistance to acceleration under an applied force. Physically, this phenomenon relates indirectly to the oscillatory differential translational velocity intrinsic to the electron or proton field, as discussed above. The greater the internal intrinsic velocity of the field (for example, the proton), the greater is the force required to achieve a specified magnitude or direction of velocity of the external field. A lesser force exerted on a field with a slower intrinsic translational velocity (for example, the electron) would produce the same effect. On the other hand, although fields with a greater $B_{\theta G}$ intensity (greater mass) exert a stronger attractive force on other fields (bodies), their acceleration toward other bodies is independent of their $B_{\theta G}$ field intensity because the enhanced attractive force due to the higher $B_{\theta G}$ field is offset by a corresponding increase in their inertia. Although the inertial effect is indirectly proportional to the magnitude of the differential translational field velocity, it is sourced from the strong interactions [23] [24] between the fibers (which are proportionate to their energies) of the electron or proton field. The gravity field and thus its gravitational potential are derived from the **net** translational movement of the fibers in the field.

As previously discussed, the intrinsic differential translation velocity of the field is proportional to the fiber/disk densities (and hence the energy) of the half-fields in the crossover zone. If the disk (or fiber) densities increase in the overall field, it automatically increases in the crossover zone, raising the inertial mass [23] [24] of the whole field. Recall that the proton field is an aggregate of electron fields and thus contains more disks/fibers (not to be regarded as additional mass) than the electron. Consequently, the proton has higher energy, inertial mass, $B_{\theta G}$ field strength, and gravitational attraction than the electron. To summarize, the intrinsic differential translation velocity allows a field to interact gravitationally with other fields. This velocity depends on the density of disks/fibers (and thus the energy) in the crossover zone. The energy in the crossover zone is increased in the proton field by the following mechanism. The fiber count increases according to Equation 1(a), imparting energy to each fiber through the compressed ∂z in the crossover zone, which raises the spatial gradient of B_θ ($\partial B_\theta / \partial z$ in Equation 1(b)). Consequently, the oscillation strokes of the fibers in the crossover zone decrease and their energies increase. The mass of the proton field is greater due to the increased energy and thus greater interaction between fibers [23] [24] in the crossover zone.

The B_θ fields resulting from the oscillatory velocity and movement of each electron (or proton) half-field along the z-axis occur simultaneously and cancel because they are equal and opposite. Therefore, such movement does not generate a gravitational field and force. Although the differential translational velocity oscillates similarly along the z-axis, it is not simultaneously generated in each half-field, and therefore does not cancel, but instead gives rise to a gravitational field.

B. Macroscopic Field: The macroscopic field is a composite of numerous microscopic fields. The microscopic, intrinsic $B_{\theta G}$ fields of the many electrons and protons comprising a body or substance, such as H_2 gas, form macroscopic alternating bands of electron- and proton-produced $B_{\theta G}$ fields (hereafter called $B_{\theta Ge}$ and $B_{\theta Gp}$ fields, respectively).

Just as electron B_θ fields accumulate and combine to form a large composite B_θ field in electrical current rather than a formation of individual B_θ fields of individual electrons, so do the bands of the $B_{\theta G}$ fields. As the $B_{\theta G}$ bands accumulate and combine within a body, such as a sphere of H_2 gas, they also grow and extend beyond the body surface. The $B_{\theta G}$ fields of larger bodies extend further because more $B_{\theta G}$ fields contribute to the overall field. The $B_{\theta G}$ field strength is greatest at the surface of the body. From here, it declines gradually to zero toward the gravitational center and also toward its baryonic limit, which is examined in Section 11.D.3.

The macroscopic $B_{\theta G}$ field bands originate from a body's electrons and protons and, like their originators, are randomly oriented. The vector summation mechanism requires that the $B_{\theta G}$ field bands in a macroscopic body realign such that their vectors do not oppose each other and can therefore coexist. This is accomplished by a spherical arrangement of the $B_{\theta G}$ field bands, where the bands are contained in the longitudinal planes cut through the origin of the sphere. The origin of each $B_{\theta G}$ field band is that of the overall spherical gravity field, which lies at the center of gravity of the body. A $B_{\theta G}$ circumferential vector occurs at each incremental radius of a $B_{\theta G}$ field band, similar the fields depicted in Figure 14. Thus, the gravity field constitutes a group of concentric spheres of $B_{\theta G}$ circumferential vectors oriented along the longitudinal lines of their respective spheres. Within each band, the longitudinal lines of the $B_{\theta G}$ field vectors intersect at the respective poles of the concentric spheres. The bands created by the $B_{\theta Ge}$ and $B_{\theta Gp}$ fields identically alternate with their individual microscopic counterparts.

Outside of a body, the gravity field is spherical, implying that the $B_{\theta Ge}$ and $B_{\theta Gp}$ field strengths are homogenously distributed at a given radius from the body's center of gravity. The gravity field is spherically distributed throughout a spherical body and is continuous with the external field; however, for non-spherical bodies, the internal and external fields abruptly change at the body surface. The $B_{\theta G}$ field of each electron and proton within a body contributes infinitesimally to each macroscopic $B_{\theta Ge}$ and $B_{\theta Gp}$ field lines, respectively.

Recall that the microscopic $B_{\theta G}$ field of each electron and proton field depicted in Figure 14 results from intrinsic and differential translational velocities of the fields. Consequently, the electron and proton fields attract via the vector summation mechanism represented by the Lorentz-type force (the gravitational force between the fields, given by Equation 13). The same principle applies at the macroscopic scale; the alternating bands of $B_{\theta Ge}$ and $B_{\theta Gp}$ fields draw the individual electron and proton fields in the body toward its center of gravity by vector summation of the overlapping $B_{\theta Ge}$ and $B_{\theta Gp}$ bands, which is similarly given by Equation 13. To calculate the gravitational force between two separate bodies, Equation 13 is applied to each body, using the sum of constants e_p and e_e, rather than the values of a single proton and electron as in a pair of interacting microscopic fields. The distance between the gravitational centers of the bodies, r_x, and v_{zp} and v_{ze}, denoting the differential translational velocities between the half-fields of the proton and electron fields, respectively, are identically treated for the microscopic case. Although this method of calculating the gravitational force between two bodies is infeasible in practice, it demonstrates the relationship between the microscopic and macroscopic gravitational interaction.

The gravitational force between two bodies arises from vector summation of the $B_{\theta Ge}$ ($B_{\theta Gp}$) field bands of the first body and the $B_{\theta Gp}$ ($B_{\theta Ge}$) field bands of the second body.[6] *If all of the bands of the two bodies are oriented in the same manner as the two fields in Figure 14, then the gravitational force is due to the vector summation mechanism (referred to above) of the y-components of the $B_{\theta G}$ field bands of the two bodies. Both are measured at the origin of the second body or vice versa. In this exercise, all of the individual microscopic field bands are equivalently assumed to originate at the origins of the two bodies and oriented per Figure 14.* If the $B_{\theta Ge}$ field bands of the first and second bodies align and the $B_{\theta Gp}$ field bands behave similarly, the bodies initially repel each other. However, the vector summation mechanism forces differential shifts in the field bands, allowing coexistence of the opposing B-vectors, and the repulsive force rapidly becomes attractive. The field bands of both bodies then oscillate in synchrony and a two-way interaction is established between the $B_{\theta Ge}$ and $B_{\theta Gp}$ field bands. Together with the intrinsic movements of the fields, this synchrony creates an attractive gravitational force between the bodies.

C. Neutron Mass: Thus far, the gravity model has been applied only to H_2 gaseous bodies, which contain electrons and protons. Higher elements contain neutrons, which while not subject to Coulomb forces *(unlike electrons and protons)*, do contribute to Lorentz-type gravitational forces. The absence of Coulomb capabilities in neutrons is attributable to their oscillating half-fields, which comprise mixtures of bidirectional B_θ fields. However, the measured mass of the neutron is comparable to that of the proton; therefore, the differential translational velocities of their combined half-fields must increase during crossover. Because the proton and neutron have roughly equal mass, the combined crossover velocity must be nearly twice that of an individual proton. Thus, the $B_{\theta G}$ field bands of protons combined with neutrons should be approximately double in strength.

Section 9 below examines the proton/neutron combination mechanism. The combined fields enhance the disk/fiber density (thus energy and mass too) in the crossover zone, increasing the infinitesimal translational velocities therein (also in the whole field).Consequently, the gravitational strength and measured mass of the combined field is greater than that in the single proton field. In this way, neutrons contribute to the overall gravitation and mass of a body.

D. Mass/Energy: In this study, the at-rest energy (Einstein's $E = mc^2$) of an electron, proton, or neutron field arises from oscillation of their radial fibers across their respective disks. As previously explained, the inertial mass of a field is also generated by the same oscillatory movements of the radial fibers of the field. This inertia must be overcome if the whole field is to accelerate, which requires sufficient force and energy. Thus, force and energy are directly related to and measured by the field's inertial mass "m," which is also a manifestation of the gravity field generated by the same oscillatory movement of the radial fibers and its net incremental translational movements.

The following is a summary of *previous examinations* with additional arguments: *As each half-field of an electron (see Figure 2) or proton contracts during an inward oscillatory movement, their innermost disks cross over at the origin and penetrate the opposite side. Immediately after one of the innermost disks has crossed the x–y plane at a certain velocity, the delayed innermost disk moves in unison at an infinitesimally greater velocity (along the z-axis). This process is responsible for the previously mentioned intrinsic differential translation velocity (v_z). The gravity field $B_{\theta G}$ is generated by the net movement of its radial fibers (measured by v_z) by which the field gains gravitational potential. From the quantity v_z, we can calculate the $B_{\theta G}$ field intensity (from Equation c) at the origin of the field, and hence, derive the gravitational potential of the field. Mass is generated by the field's energy which is derived from the oscillation strokes of its fibers.* Because mass is related to v_z since gravitational force is per Equation 13, the internal at-rest energy is related also. *Recall that in the configuration of Figure 14, the gravitation and thus, mass of either field are based on or related to its $B_{\theta G}$ field contributed by all of the field's radial fibers;* hence, the internal at-rest energy must also relate to the $B_{\theta G}$ field since mass does. *Furthermore, recall that the B_θ intensities are computed as the sum of B_θ increments (to the point of measurement) induced by the radial fibers oscillating across and twirling around their disks;* thus, gravity and mass can be regarded as arising from a single representative fiber in the crossover zone, oscillating across the x–y plane along the z-axis with differential velocity $\pm v_z$. Referring to Figure 14, this fiber is equivalent to the sum of all radial fibers creating the $B_{\theta G}$ intensities at the y-tangents (*or the sum of the incremental B_{YG} components from theoretical plus and minus infinity to the point of measurement at the virtual origin*). Consequently, the at-rest energy of the field equals the kinetic energy of the representative radial fiber since the inertial mass of the field is given by this fiber. The inertia and energy of the representative fiber is the sum of the inertias and energies, respectively, of all radial fibers in the field, since all of which contribute to the $B_{\theta G}$ field. The frequency of the representative fiber is the energy of the at-rest field (mc^2) divided by Planck's constant. The representative fiber is identifiable only as the fiber in the delayed innermost disk at the instant of reversal of the half-fields. All other fibers may be completely stationary, their energies absorbed by the representative fiber, whose oscillation stroke would significantly shorten as the B-elements of the other fibers faded.

Because the proton is more massive than the electron, its at-rest energy is also greater. This mass disparity can be understood as follows. As previously explained, the proton field is a composite of electron fields, and thus contains a greater density of disks/fibers than the electron field. In the crossover zone, this increased density (and energy) increases the inertial mass and the intrinsic differential translation velocity (v_z) of the field. The $B_{\theta G}$ field strength, gravitational attraction and internal at-rest energy are all greater for the proton than the electron. The same physical processes apply when a proton is combined with a neutron as well as in macroscopic bodies, which constitute a bulk of electron, proton, and neutron fields. The oscillation strokes of proton fibers are shorter than those of the electron, particularly near the origin where the overall disk/fiber density is relatively high. Thus, the oscillation frequencies, fiber energy (Σhv) and v_z are all elevated in the proton, leading to higher mass and greater rest energy (mc^2).

The measured masses of electron and proton fields and the internal energies of the fields are related to the density of the disks/fibers within the field, and therefore, vary with fiber density. For example, the number of disks/fibers in a field can change by radiation emission/admission (not to be regarded as mass) or by increased/decreased binding strength to another field. For example, as two fields become more tightly bound, fibers may be emitted, thereby reducing the mass, internal energy, and intrinsic differential translation velocity of the field due to the decreased fiber count. The internal at-rest energy obeys mc^2 but with smaller m.

Because fields such as electrons and protons translate at some velocity, they possess external kinetic energy as well as internal rest energy. The external energy is calculated as $(\gamma - 1)mc^2$ and the total energy is γmc^2. The external kinetic energy affects the internal mechanism of the field through the addition of more fibers upon acceleration of the field. The additional fibers increase the fiber density in the crossover zone, and therefore increase the differential translation velocity of the field by the mechanism described earlier. The consequences are greater mass inertia and a correspondingly stronger $B_{\theta G}$ field and gravitational attraction, relative to the at-rest field. The field's mass increases from m to γm, while the total energy remains at

γmc^2. Although the well-known net B_θ field of the electron also arises from whole-field translation, this field cannot interact with the gravity field of another body because the B_θ field induced by translation is homogenous and unidirectional, while the gravity field consists of alternating $B_{\theta Ge}$ and $B_{\theta Gp}$ field bands.

9. Nuclear Fusion and Other Elements

As is well documented, large volumes of hydrogen gas aggregate into stellar formations under gravitational attraction between the atoms in the gas *(more specifically, between the gravitational fields of the electrons and protons in the context of this study)*. Although the core of the star is not the site of its strongest gravitational force, the core is under the highest gravitational pressure being subject to the accumulation of all atomic gravitational interactions between the center of the star and its core radius, and beyond the core to the star's periphery. Not only do interacting gravitational fields and forces accumulate toward the center but also toward reduced spherical surface areas to resist the accumulating forces.

The innermost parts of developing stars experience the greatest gravitational pressure. Once the star has grown sufficiently for this pressure to overcome the Coulomb barrier, the gaseous structure [depicted in Figure 13(b)] of its inner part collapses into plasma. As the star attracts more gas and increases in size, the sphere of inner plasma increases similarly. A star of adequate size creates enough pressure in its core to fuse the hydrogen plasma into helium. Associated with this pressure is a greater density of electrons, protons, and their emitted radiation in the core, which raises its temperature. The radiation released by core fusion processes heats the outer regions of the star *(due to admission of the escaping radial heat fibers)*, thereby converting these regions to plasma as well.

As mentioned above, hydrogen plasma under extreme pressure can fuse into helium. Although this pressure can dissociate the hydrogen gas structure into a plasma of free electrons and protons, it can fuse neither two protons nor two electrons. It is much easier to fuse a closely spaced electron and proton as shown in Figure 8(a), except that one of the electron fields is replaced by a proton field. Under the required pressure, these two fields are merged into a single field—the neutron. The B_θ half-fields of the neutron field are neutral *(uncharged)* because the B_θ half-fields of the electron oppose those of the proton. These opposing fields are responsible for the neutron's tendency to revert to separate electron and proton fields; this mechanism is popularly known as the weak interaction. Effectively, the neutron owes its creation and stability to the gravitational pressure it is under at this stage in its development.

The gravitational pressure initially forces the opposing B_θ fields of the neutron to coexist; however, as the neutron field is again forcibly interacted with free protons and electrons by the gravitational pressure, these fields become further locked in. These interactions also occur side by side, resulting in formation of helium atoms, as shown in Figure 15. The half-fields of the protons and neutrons comprising the nucleus oscillate in unison along their z-axes. Each of the two electrons partially interacts with both protons. The helium gas is structured similarly to hydrogen gas, except that it is monatomic rather than diatomic (as in Figure 13).

The two proton fields in the nucleus of the helium atom can now coexist because two neutron fields are present. The neutrons buffer the protons because they are closer to and directly interact with each proton. Their shared neutrons prevent the protons from directly interacting with each other, permitting their coexistence in a single nucleus.

Under gravitational pressure, the right-handed B_θ half-fields of each proton are forcibly interacted with the attractive left-handed B_θ half-field components of each neutron field. They are also forced to coexist and entangle with the right-handed B_θ half-field components of the neutrons. This mechanism not only results in non-uniform movements and uneven distributions of the proton and neutron fields but also reinforces their interconnection, giving rise to the strong nuclear force. This connection is retained even in elements that are free of stellar gravitational pressure. These interactions are mirrored for the left-handed electron B_θ half-fields shown in Figure 15, further strengthening the binding of the atom constituents. The electron half-fields oscillate outward as the proton and neutron half-fields oscillate inward and *vice versa*, inducing two attractive forces (one a partial Coulomb force, the other a Lorentz force) between the overlapped sections of the electron and proton fields. These forces also contribute to binding the atomic components. The strongest bonds form when proton and neutron fields combine into a composite field, which then connects to a pair of electrons.

As stated above, the hydrogen plasma generated in the stellar core is under extreme gravitational pressure, and is therefore extremely dense. Such high pressure and density compresses the cylindrical fields of the electrons and protons of the plasma, forcing them to emit some of their radial heat fibers (photons) as discussed in previous sections. These fibers either exit the star or are captured by stellar electrons and protons more distant from the emission point. The enhanced density of radiated fibers within the star's core increases the plasma temperature, although the fibers have no intrinsic temperature. The relationship between fiber density and temperature is examined in Section 11. When the hydrogen plasma fuses into helium, more radial heat fibers are emitted from the cylindrical fields of the electrons, protons, and neutrons as the fields become more tightly bound.

Other elements in stars and supernovae are formed similarly to helium; this process has been extensively documented. As the gravitational pressure in larger stars (or supernova pressure) increases, more protons and electrons are combined into neutrons *(in the context of this study)*. The neutron fields become fused with other proton fields, and also to previously-fused elements, thereby forming new elements with greater nucleon counts and masses.

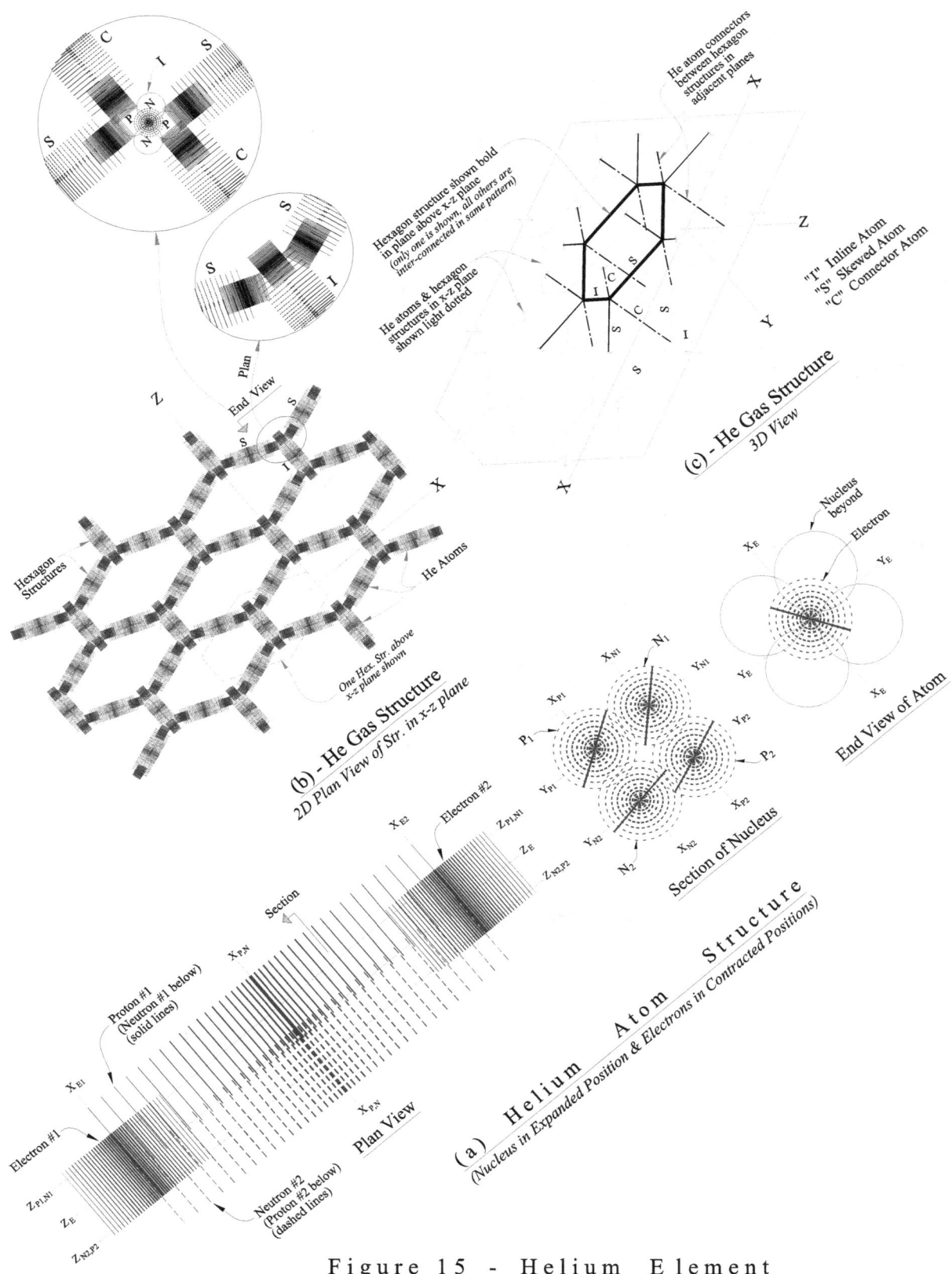

Figure 15 - Helium Element

Neutrons must exist between most of the protons to buffer their opposition. The interactions and connections between the nucleon fields occur under extreme pressure, which enforces and trains them to coexist. Each element is characterized by a different arrangement of its nucleon and electron fields. To secure a more even distribution of their positive members, the nucleons in the nuclei of other elements may be combined, have additional parallel arrangements or may form a population of parallel and crisscrossing individuals.

10. Hadrons, Leptons, Bosons and Neutrinos

In this study, protons and neutrons *(collectively labeled as hadrons)* and electrons *(labeled as leptons)* are the only B_θ fields commonly found in nature. Photons, which are similarly ubiquitous in nature, constitute a series of oscillating radial fibers *(labeled as bosons)* emitted from any of the three common B_θ fields. Neutrinos, which are measured as fermions, must also be composed of B_θ field because B_θ fields are fermions (as previously discussed) and oscillating radial fibers emitted from B_θ fields are bosons. Because a neutrino is inferred to be very light with no measurable charge, it may be a miniscule fragment of a neutron B_θ field detached from its parent during a nuclear reaction or decay. The photon and neutrino fields are configured such that they do not generate gravitational $B_{\theta G}$ fields (see Section 8; "Gravity"), but they can respond to gravitational fields via movement of their fibers relative to a $B_{\theta G}$ field.

A plethora of hadron and lepton particles have been observed or inferred in special events such as collisions in particle accelerators. During such an event, a B_θ field (for example, a proton) collides at speeds sufficient to compress further its already Lorentz-contracted radial heat fibers/disks into an extremely dense state. Supposing that the colliding fields oppose each other, the densities and speeds of their radial fibers/disks at the contact point are so great that the fields cannot intercept without losing and scattering some of their fibers/disks. Recall that both half B_θ fields of an electron, for example, cross to the opposing side as they oscillate inward. In this case, the half-fields pass through/by each other in an orderly fashion. If the B_θ fields collide at great speeds and densities, this smooth transition is no longer possible.

The dissociated and scattered fibers/disks emanating from high-impact B_θ field collisions can appear as radiation or as numerous uncommon hadron and lepton B_θ fields. These B_θ fields instantly form by vector summation *(previously examined)* of the scattered fibers/disks and decay equally rapidly because they have likely formed ad hoc from randomly oriented fibers ejected at high speed. Because the fibers may oppose each other and/or are desynchronized, they either segregate and appear as radiation or the unstable B_θ field decays into a stable B_θ field such as a proton. The decay process is similar to that of a neutron field to a proton field as previously examined; however, multiple transitions to more than one type of B_θ field may occur before a stable B_θ field is obtained or the fibers completely segregate and escape as radiation.

The mass measurements of the numerous uncommon B_θ fields described above vary widely. As examined in the section "Gravity," the magnitude of a gravitational $B_{\theta G}$ field derived from a B_θ field *(thus, its mass measurement)* arises from the net crossover velocity of its half-fields at its origin. This velocity depends chiefly on the density of the fibers/disks in the crossover zone of the B_θ field. Because these parameters can vary widely in the ad hoc B_θ fields created from high-impact collisions, their mass measurements, as well as their charge measurements, will vary similarly. The components of neutral B_θ fields have opposing circumferential B-vector directions much like the neutron previously examined. By contrast, the components of composite B_θ fields with plus or minus integer values of e rotate in the same circumferential direction. Because all of these fields are unstable as described above, they either revert to an electron or proton field *(or their anti-fields with opposite circumferential B-vector directions)* or disassemble into radiation.

The transverse spread of oscillating heat fibers, referred to as quark-gluon plasma, has been witnessed in some heavy nuclei colliding at light-like speeds in particle accelerators. This plasma is the *first phase of matter* resulting from the collision. Although the individual heat fibers possess no temperature, the plasma is extremely hot because the colliding fibers are extremely dense and energetic (the relationship between fiber density/energy and temperature is examined in Section 11). Instantly, these fibers either coalesce into B_θ fields *(second phase of matter)* by the vector summation mechanism or individually radiate out. In this study, the vector summation mechanism is the "strong interaction" that forms the B_θ fields; in conventional theory, quark particles are bound into hadrons by a mechanism called the "strong interaction", analogous in name only. Recall that hadrons are B_θ fields in this study. The observed collinear rotational movements of the hot plasma as it converts to B_θ fields are likely due to the collective formations of the B_θ fields: fields with left- and right-handed rotations move together in different directions. For a thorough discussion on this topic, see here. [22]

11. Closing Examinations

This study proposes the oscillating heat fiber or photon (Figure 1) introduced in Section 1 as the fundamental component of matter. A "heat fiber" [7A] is characterized by three fundamental intrinsic properties: energy, oscillatory motion with or without twirling, and inclination to join and coexist with other fibers via the vector summation mechanism. This mechanism, which is the primary driver of all field creations (matter) and interactions studied in this work, apparently derives from the tendency for fibers to move toward the perpendicular part(s) or B-element(s) of nearby fibers with similar rotational directions; if the directions are dissimilar, the fibers tend to separate or move apart. This mechanism is consistent with Maxwell's Equations.

The kinetic energy of a photon (heat fiber), given by Planck's law $E = h\nu$, arises from its light-speed oscillatory motion rather than its translational motion. The energy of this oscillation is related to the oscillation range; shorter ranges yield shorter wavelengths and higher frequencies, and thus greater energies. The fiber energy $h\nu$ is the limit of the external work that may be performed by a fiber during an interaction with a system or the internal energy contribution of a fiber to a body.

A. Energy: The fiber's intrinsic perpetual oscillating motion gives rise to the laws of energy conservation; that is, without such motion, energy would not be conserved in any process. A fiber's energy may vary during interactions with other fibers or as it contributes to the internal energy of a stationary body. For example, a fiber that becomes part of an electron B-field disk may alter its oscillation range. As explained in Section 2, fibers/disks nearer the x–y plane of the electron are closer together, and thus denser, than more distant fibers/disks. The higher fiber densities ensure smaller effective disk diameters with short oscillation ranges and consequently higher energies. The increased energy of fibers near the x–y plane manifests as greater external kinetic energy upon their emission from the electron field. The opposite effect yields smaller energies for fibers at both tail ends of the electron, where disks are wider and less dense. Hence, the mechanism of the oscillating electron field depicted in Figure 3 accounts for the varying energies of the emitted fibers. Summing the energies of all fibers in the electron, we obtain the sum of the kinetic energies of the same fibers prior to their joining the electron field. The energies of the fibers in the B-fields vary similarly in the electrons, protons, and neutrons of atoms and molecules.

B. Momentum: Momentum transfer between macroscopic bodies occurs via transfer of fibers/disks between the bodies as they collide. The translational velocity of the macroscopic body is the vector sum of the translational velocities of its composite B-fields, as examined previously. A simple example of momentum transfer follows: as a moving body collides head-on with a less massive stationary body, some of its fibers/disks are transferred (impressed) into the stationary body; the transferred fibers/disks join and simultaneously mobilize the fibers/disks of the B-fields comprising the stationary body, jostling them in varying speeds and directions. The resultant sum of these individual B-field movements enables the motion of the stationary macroscopic body in the direction of the moving body.

After the collision and the at-rest body is set in motion, some of the transferred fibers/disks are returned from the B-fields of this body back to the first moving body; this opposes the resultant movement of its B-fields and thus the momentum of the first moving body decreases accordingly. Conservation of momentum of the system is possible due to the proportionate sharing of these fibers/disks between the macroscopic bodies. The at-rest body is inclined to return the transferred fibers/disks back to the moving body since its B-fields are taken out of equilibrium and are required to assimilate the additional fibers/disks; but it is only able to do so to the extent the B-fields of the moving body allow. The proportion of the fibers/disks that are returned is related to the fiber/disk counts and velocities of both bodies. As examined in the section under *Gravity*, the mass measurement of a body is proportional to its fiber/disk count. Thus, a more massive moving body will undergo less change in its velocity since it would provide more opposition to the returning disks/fibers and thus take in less of them; and further, it contains more disks/fibers that have to undergo a velocity change. This allows the conservation of momentum law: $\Delta m\mathbf{v}$ for each body is equal and opposite to that of the other body after an elastic collision (for example, all fibers are collectively retained in the two bodies and none is radiated). If the at-rest body were more massive than the moving body, the fibers/disks transfer mechanism would be similar which may cause the moving body to reverse its direction.

A single incoming heat fiber (photon) can increase the momentum of a macroscopic body (or an individual B-Field) despite being massless. A photon lacks "mass" and gravity field because both quantities are related to the incremental translational oscillatory movements of a B-field along its z-axis during crossover, as explained in the *Gravity* section. Macroscopic bodies possess mass and gravity fields because they comprise B-fields made up of individual heat fibers. Nonetheless, a heat fiber can increase a body's momentum by attracting (vector summation) a disk component(s) of one of the B-fields of the macroscopic body as it passes; it can also be admitted between two disks (if resonant with the disks). This interaction minutely adjusts the translational velocities of the disks, B-fields, and ultimately the macroscopic body. A high-energy heat fiber imparts greater momentum to a body because it impacts more heavily on the B-field disk components due to its higher frequency.

Other phenomena that may result from interaction between an incoming photon and a macroscopic body or an individual B-field are as follows: 1) the photon may deflect from one of the fibers in the body with less, the same, or more energy depending on the frequency (energy) of the deflecting fiber; and 2) a high-energy incoming photon fiber may interact with a single B-field, imparting enough energy to eject the field with scattering of a less-energetic photon; this phenomenon is popularly known as *Compton Scattering*.

C. Temperature: Each "heat fiber" has the same intrinsic energy content [7A]. Fiber energy is governed by the length of its oscillation stroke, not by the volume of its energy content; thus, the energy levels of fibers are allowed to vary despite the constant intrinsic energy content. Our sense and measurement of temperature[7B] are determined by the density and energy of free heat fibers at the point of interest.[8] If the density of free fibers is enhanced at a particular location, then the measured temperature at this location is higher than that at a less dense location. If the density of free fibers is the same at two locations but the fibers are more energetic at one location than the other, then the temperature is also higher at that location.

Free heat fibers are not bound to a B-field; fibers bound to B-fields usually reside in macroscopic bodies. In general, free heat fibers include all types of electromagnetic radiation. The measured temperature of a body measures the density and energy of the heat fibers radiated from the body. When taking a measurement, the B-fields of a temperature tool are excited to calibrated levels by admission of the radiation ejected by the measured body; the denser and/or more energetic the radiation, the more excited its B-fields, and the higher the registered temperature. A body emitting no radiation yields no temperature measurement.

D. Cosmology: As presented in this study, matter is composed of one or more cylindrical formation(s) of groups of oscillating and twirling heat fibers. The cylindrical formation (see Figure 2) has been referred to generally as a B-field, or more specifically as a B_θ field. As matter formed in any nascent region of the universe, only a very dense quantity of oscillating heat fibers (such as shown in Figure 1) existed. On the basis of our notion of hot and cold, the temperature in the nascent region was extremely high owing to the high density of heat fibers.[7] As explained above under *Temperature*, the measured temperature would have decreased as the free fibers became bound into the forming B-fields. Speculating, the heat fibers had moved from outside into our universal space from all directions. As the fiber density increased in an initial part of this space, the fibers began to form cylindrical structures; such development continued, extending in all directions from the initial zone, while fibers continued to enter the development zone from outside. Today, the development of cylindrical formations has either ceased or the development zone continues to function beyond the observable universe.

D.1. B_θ Field Structure: An intrinsic property of a heat fiber is its inclination to join other heat fibers through the vector summation mechanism. When their oscillatory movements and also the movements of their perpendicular parts oppose similar movements in other fibers, they separate rather than join. The forming and disjoining of their perpendicular parts during oscillations allow the fibers to remain separate when connecting with other fibers, for example, during the formation of a B-field. The cylinder (see Figure 2) is the only permissible structure of oscillating heat fibers, for the following reasons: 1) As oscillating fibers join in the vector summation mechanism, they are drawn towards the perpendicular parts (B-elements) of neighboring fibers. 2) To enable this process, the alignments of the fibers continually adjust, which gives way to synchronized twirling of the fibers and alignment of the fiber origins (Figure 1) along the z-axis of the cylindrical field per Figure 2. 3) The twirling fibers can only coexist in the resulting cylindrical formation such that any one fiber will not oppose any other fiber as the fibers oscillate across and twirl in their respective disks, and as the cylindrical field oscillates back and forth along its z-axis. 4) The lengths of the cylindrical group of fibers are limited by the mutual attraction of them in the z-direction of the field; a group that is long would separate into individual smaller groups. The above fiber interactions and behavior were substantiated using Maxwell Equations in Section 2.

The cylindrical B-field can deflect a non-resonant incoming fiber (a photon that is out of phase or within a very different oscillation range from the intercepted fibers). Fibers that oppose the direction of fiber movement in the field or those that attempt to enter an already crowded field may also be deflected. The B-field can withstand such local disturbances from incoming photon(s) while maintaining its cylindrical structure. An intruding fiber oriented parallel to the z-axis passes through the field without disturbing the perpendicular cylindrical fibers. The re-emergence of various B-fields immediately following collisions in particle accelerators exemplifies the original cosmic formation of the B-fields under the vector summation mechanism, but the re-emergence occurs with greater fiber count density.

D.2. CMB Radiation: The mainly isotropic CMB radiation pervading our known universal space is likely a residual mix of heat fibers remaining from the cosmic formation of the B-fields out of the extremely dense heat fibers originally pervading universal space. The present-day CMB radiation is not dense enough to sustain B-field development. Because of their black body type distribution, the CMB heat fibers probably resided in B-fields at some stage, and were subsequently emitted at their corresponding energy. Alternatively, they may have combined with residual parent fibers. The long elapsed time between the formation and observation of the CMB has allowed its constituents to spread, fill voids left from matter creation, commingle, and generally redistribute into an isotropic mix. The minuscule temperature anisotropy of the CMB radiation and its spatial distribution pattern could arise from a number of factors. For instance, the original dense parent radiation may not have been perfectly isotropic and may be preserved in the current CMB. Alternatively, CMB radiation fibers with similar oscillation frequencies may preferentially populate specific areas or zones, such that resonant fibers tend to cluster in regions separate from their non-resonant counterparts.

The formation of hydrogen from the B-fields, stars from hydrogen gas, and more complex matter from stars was investigated in previous sections of this study. The starlight observed from very distant galaxies is cosmologically red-shifted. The more distant the galaxy, the more its starlight is red-shifted. In the new framework, cosmological redshift is attributable to numerous minute interactions between the starlight's oscillating heat fibers (photons) and other oscillating heat fibers (photons) that intercept the paths of the traveling starlight's fibers. The energy lost to each interaction is miniscule but when light travels over extremely long distances, it undergoes sufficient interactions to lose significant energy, giving rise to the measured redshift. A rough calculation will illustrate this effect:

As reported, a very distant galaxy was roughly measured at 13 billion light-years from Earth or 1.23×10^{26} m at the time it emitted its light. If the frequency of the emitted light is 6×10^{14} Hz (oscillation range = 2.5×10^{-7}m) with a corresponding energy of 2.48 eV and a measured redshift z ≈ 7.0, then the observed frequency is 7.49×10^{13} Hz (oscillation range = 2.0×10^{-6}m) with a corresponding energy of 0.31 eV. The total energy decrease of a single oscillating heat fiber in the starlight over a distance of 1.23×10^{26}m is the difference between these energies, 2.17 eV.

The radiation photons (fibers) most likely encountered by the starlight photons (fibers) are CMB photons. Hence, we consider only CMB photons, although the starlight could intercept photons from sources, such as other starlight. The mean wavelength and frequency of CMB radiation is 1.76×10^{-3} m and 1.70×10^{11} Hz (i.e., oscillation range = 8.8×10^{-4} m), respectively, giving a mean energy of 7.05×10^{-4} eV. Mean fiber density of CMB is 3.78×10^{8} fibers/m³.

Because the radiation fibers oscillate at near-light speed, the oscillation range given above assumes c as the oscillation speed. The interaction, in this example, is the vector summation mechanism as two oscillating fibers move relative to each other along a common z-axis. The movement of a fiber (photon) is shown in Figure 10. As the fiber translates along the z-axis, it oscillates in planes perpendicular to the z-axis: up and down movements along its own axis give rise to linear movements during translation as shown; alternatively, the fiber can move back and forth along its own axis while rotating in the same planes, yielding circular movements during translation. In this example, both starlight and CMB photons are assumed to translate while rotating in a circular fashion.

Maxwell's Equation 1(b) given in Section 2 depicts the vector summation mechanism of the above-described interaction between the two oscillating fibers. The average diameter of the circular cross section of the light fiber path equals its average oscillation range, 1.125×10^{-6}m. This cross section is smaller than the circular cross section of the CMB fiber path whose diameter is 8.8×10^{-4}m. If two concentric circular cross sections (the inner and outer cross sections representing the light fiber and CMB fiber paths, respectively) translate along a common z-axis, the perpendicular parts (B-elements) of the CMB fiber will reside on average outside the perpendicular parts (B-elements) of the light fiber. The two fibers can coexist in this configuration if their rotational directions are the same: two fibers translating in opposite *directions must rotate in opposite directions, while two fibers translating in the same direction must rotate in the same direction (either right-handedly or left-handedly). Other combinations cannot coexist, thus interactions derived from these combinations also do not occur. Similar arguments apply to combinations of fibers with the same rotational direction but whose origins (Figure 1) do not align or nearly align along a common z-axis.*

Equation 1(b) describes the interaction between a light fiber and CMB fiber as they translate relative to each other along a common z-axis in the above configuration. Physically, the radial light fiber is drawn to a B-element(s) of the radial CMB fiber as the two entities pass each other and vice versa. *This combination imposes an attractive force (E_h in Equation 1b) between the B-elements of both fibers. Because, on average, the B-elements of the CMB fiber reside outside the elements of the light fiber, this force extends the oscillation stroke of the light fiber while shortening that of the CMB fiber, thereby decreasing the energy of the light fiber and increasing the energy of the CMB fiber. If the relative velocity between the fibers, v_z in Equation 1(b), is zero, then the force between the B-elements fails to mobilize and no energy is transferred.*

The volume of space within which an individual starlight fiber may interact with a CMB fiber throughout its journey equals the cross-sectional area of the origin of the light fiber times its length of travel. In this study, the diameter of the field origin is conservatively set as 2 fm. For this example, we assume that only CMB fibers whose origins (Figure 1) occur within the origin space of the light fiber can interact effectively with the light fiber in the above-presented way. The cross-sectional area of the light fiber origin is $3.14 \times 10^{-30}\,m^2$. Hence, the volume of space in which an individual starlight fiber can effectively interact with a CMB fiber is $3.14 \times 10^{-30}\,m^2 \times 1.23 \times 10^{26}\,m = 3.86 \times 10^{-4}\,m^3$.

The number of CMB radiation fibers existing in this volume of space at any instant is their density = 3.78×10^8 fibers/m³ multiplied by $3.86 \times 10^{-4}\,m^3$ i.e., 1.46×10^5 fibers. Half of these fibers rotate oppositely to the light fiber and thus do not participate in the interaction configuration. The remaining half (7.3×10^4 fibers) can interact with the light fiber according to Equation 1(b). Each oscillating and rotating CMB fiber creates an effective circular magnetic field B_θ whose intensity decreases with increasing radial distance from the fiber origin. The rotating light fiber generates a similar B_θ field of greater intensity because the oscillation strokes of this fiber are shorter. In each of the 7.3×10^4 intercepts of a CMB B_θ field with a light B_θ field, a minute amount of kinetic energy is transferred. The average kinetic energy transferred per interaction is 2.17 eV divided by 7.3×10^4 interactions = 3×10^{-5} eV. CMB fibers, residing outside the circular zone of the light fiber as described above, have offsetting interactions with the subject light fiber and thus, no net energy is transferred during these interactions.

As examined in Section 2, the interaction between the two B_θ half-fields of the electron [the vector summation mechanism of (Equation 1a)] causes the half-fields (Figure 2) to oscillate along the z-axis. The force E_z that drives this oscillation is similarly derived to the E_h force derived from the two-photon fields in the above configuration. Mobilization of the E_z type forces, such as those causing an inward oscillatory movement, arise from the attraction of a radial disk fiber to the perpendicular parts (B-elements) of a radial fiber in an adjacent forward disk; the opposite effect occurs in the overlapped regions of half-fields, yielding an outward oscillatory movement of the half-fields.

Similar to the E_h force derived from the two-photon fields in the above configuration, E_h type forces are also mobilized in the electron's B_θ half-fields as they oscillate along the z-axis. This was examined in Section 2 as follows: As the B-disks of each half-field transit toward the x–y plane, the effective diameter D of each half-cylindrical field gradually decreases to a minimum at the x–y plane. This shape arises from the increasing concentration of B-elements within the inner zone. The outer disks also approach and accumulate near the inner zone. The high concentration of perpendicular fiber components (B-elements) in the inner zone (particularly near the z-axis) attracts fibers in rear disks, whose oscillation ranges accordingly decrease. Thus, the diameter of each half-field is minimized at the x–y plane. The vector summation of this process is given by Equation 1(b). Because all fibers continue to oscillate at near-light speed during transit of their half-fields, their oscillation frequencies and ranges increase and decrease, respectively, as the disks in each half-field approach the x–y plane. As their frequencies increase, the fibers gain energy. The B_θ field of the CMB photon is analogous to the B_θ field of an outer disk, whose density is increased due to an increasing spatial field gradient $\partial B_\theta / \partial z$ induced by the above process. However, the density of the light photonic B_θ field, in the example, decreases because the photon lacks an adjacent denser disk, as it would have if it was in an electron field.

Continuing the above example, the B_θ fields of both light and CMB photons or fibers are obtainable from the modeled B_θ half-fields of the electron in Section 2 and Equations (c') and (c''). The reader should also refer to Figure 2. Substituting z = 1 fm and v_z = 0.099c (see Figure 4) into Equation (c') and inserting the result into Equation (c''), the effective intensity of the electron B_θ half-field is 6.10×10^{10} T. Because the oscillating x-ray fiber (photon) is the most energetic fiber in the electron field, the disks containing these fibers must occur at or near z = 1 fm. Hence, a median x-ray fiber located at z = 1 fm has not only a B_θ intensity of 6.10×10^{10} T (the B_θ intensity of the electron's half-field, being measured at z = 1 fm, is the B_θ intensity of the x-ray fiber disk, contributed by the ΣB_θ increments of the half-field) but also a median oscillation range of 4×10^{-10} m. The latter calculation assumes an approximate median x-ray frequency of 3.75×10^{17} Hz and an oscillation speed of c. Using a scale derived in Section 6 that was based on the oscillation range of the x-ray fiber and its location (z = 1 fm), we can roughly locate the disks containing the light fibers and the CMB fibers at their respective positions on the z-axis within the electron B_θ field. The oscillation ranges are 1.125×10^{-6} m and 8.82×10^{-4} m for the light and CMB fibers, respectively. In Section 6, the following approximate scale was established: $z^3 = 6.25 \times 10^{-27} D^2$, where D = the fiber oscillation range or stroke; also the disk diameter. This calculation yields z = 2.05×10^{-13} m and 1.73×10^{-11} m for the light and CMB fibers, respectively. Substituting z and v_z obtained from Equation 6 into Equation (c') and inserting the result into Equation (c''), the effective B_θ intensities are estimated as 1.46×10^6 T and 2.57×10^2 T for the light and CMB fibers, respectively.

Because these effective B_θ intensities were obtained from the average oscillation ranges of the light and CMB fibers, the average E_h force imposed on a circular twirling light fiber field by the passing of a circular twirling CMB fiber field, is calculated by substituting $\partial B / \partial z$ in Equation 1(b) with 2.57×10^2 T $\times \cos 52°$, where the weighted average angle of incidence is approximately 52°. At any weighted angle of incidence, the average v_z in Equation 1(b) is c *because an equal proportion of CMB fibers move in the same and opposite directions as the light fiber. E_h is then calculated as 4.78×10^{10} N/C or 7.658×10^{-9} N/electron. To convert N/electron to N per light fiber, the force is multiplied by the ratio of the effective B_θ intensity of the light fiber (1.46×10^6 T) to that of the electron half-field (6.10×10^{10} T, as calculated above); this is appropriate because the B_θ intensities drive the electric forces of these fields. The converted result is 1.83×10^{-13} N/γ. Multiplying this average force with the overall increase in the length of the oscillation stroke of the light fiber as it transits to earth (1.75×10^{-6} m), the energy loss of the light fiber is obtained as 3.20×10^{-19} J. This loss converts to 2.00 eV, slightly less than the measured energy decrease of 2.17 eV.*

Given the large number of variables involved in both calculating and measuring these values, such close agreement is unexpected. This example assumes a uniform CMB distribution (based on present measurements) over the entire traveling distance of the light photon. This assumption may be invalid because the intensity and density of the CMB radiation may have decreased since the time the light was emitted. Thus, more light energy would have been lost at the beginning of the photon's journey than at the end. The average force imposed on the light fiber by the passing CMB fiber (and vice versa) was calculated as 1.83×10^{-13} N/γ. During its transit, the light fiber encounters this force multiple times; the change in the length of the oscillation range of the light fiber during an average interaction is very much smaller than its overall change (1.75×10^{-6} m). Because the force applied in each interaction is instantaneous, it must exert a very small fractional effect on the light fiber; but the accumulation of such interactions results in a measurable energy loss of 2.17eV, as mentioned above.

D.3. Galactic Gravity: Recall from Section 8B that the macroscopic $B_{\theta G}$ field bands originate from the electrons and protons of a body and, like their sources, are oriented in random directions. The vector summation mechanism requires that the $B_{\theta G}$ field bands of a macroscopic body realign into positions and orientations that remove the opposition of their vectors and secure their coexistence. This is accomplished when all of the $B_{\theta G}$ field bands combine into a sphere; the $B_{\theta G}$ field bands are contained in the longitudinal planes that intercept the origin of the sphere. The origins of each $B_{\theta G}$ field band coincide with that of the overall spherical gravity field at the body's center of gravity. A $B_{\theta G}$ circumferential vector occurs at each incremental radius of a $B_{\theta G}$ field band, similar to one of the fields of Figure 14; thus, the gravity field is a group of concentric spheres of $B_{\theta G}$ circumferential vectors located and oriented along the longitudinal lines of their respective spheres. The longitudinal lines of the $B_{\theta G}$ field vectors within each band intersect at the respective poles of their concentric spheres. The $B_{\theta Ge}$ and $B_{\theta Gp}$ field bands alternate identically to their individual microscopic counterparts. Because the proton $B_{\theta Gp}$ field bands of the body are strengthened by coexisting neutron B_θ fields, its measured mass and inertia is predominately contributed by the combined proton and neutron B_θ fields, as examined in Section 8C.

Outside of a body, the spherical gravity field ensures homogenous distribution of the $B_{\theta Ge}$ and $B_{\theta Gp}$ field strengths at any given radius from the body's center of gravity.[13] If the body is spherical, the gravity field is spherically distributed throughout the body and interfaces evenly with the outside field. For non-spherical bodies, however, the inner field changes abruptly at the body surface. The $B_{\theta G}$ fields of each electron and proton within a body contribute equally and infinitesimally to each of the macroscopic $B_{\theta Ge}$ and $B_{\theta Gp}$ field lines of the body, respectively.

The effective range (or diameter) of the spherical gravity field of a body or group of bodies is limited by and related to the number of protons contained therein; the range is also limited by the oscillation range of proton fibers in the crossover zone of their B_θ fields, since it is much less than that of electrons (see Section 8A for meaning of the crossover zone and the source of gravity fields). Hence, the range of the $B_{\theta Ge}$ fields of a body or group of bodies imposed by the electrons is much greater than that imposed by protons.

The most energetic fibers emitted from electron and proton B_θ fields are x-rays (~3.75 × 10^{17} Hz) and γ-rays (~7.5 × 10^{19} Hz), respectively. Given that the fibers oscillate at or near the speed of light, the mean fiber oscillation range (or mean diameter of the B_θ field at or near the crossover zone, from where the fibers that give rise to gravity fields occur) is 4 × 10^{-10} m and 2 × 10^{-13} m for the electron and proton, respectively. Because the oscillation range for the proton $B_{\theta G}$ field is less, this field controls the range of the macroscopic gravity field.

The reported estimated baryonic mass of the Milky Way galaxy, which comprises approximately 76% hydrogen, is 1.5 × 10^{41} kg. Thus, the approximate number of protons in the Milky Way is 8 × 10^{67}, and the effective diameter of its spherical gravity field, contributed by its baryonic B_θ fields, is approximately $[8 \times 10^{67} \times (2.0 \times 10^{-13} \text{ m})^2]^{1/2} = 1.8 \times 10^{21}$ m or 1.9 × 10^5 ly. Beyond this spherical zone (called the baryonic gravity zone), radiation fibers or photons such as CMB fibers may couple with the fibers and B-elements of the $B_{\theta Gp}$ field lines through the vector summation mechanism (Equation 1b in Section 2). Such coupling would extend the effective gravity field beyond the range of the effective baryonic gravity zone.

The coupling mechanism is similar to that occurring in interactions between the CMB photons and traveling light photons explored in Subsection D.2. Outside of the baryonic gravity zone, the radial fibers of the $B_{\theta Gp}$ field lines are attracted to the perpendicular B-elements of the CMB fibers (and vice versa). Throughout these interactions (Equation 1b), the CMB fibers couple to the radial $B_{\theta Gp}$ field lines and become extensions of them. Within the baryonic gravity zone of a group of bodies such as the Milky Way, the oscillation ranges of the microscopic $B_{\theta Gp}$ field lines (from which the macroscopic $B_{\theta Gp}$ field lines are derived) are controlled by the B_θ fields of the emanating protons; thus, the proton B_θ fields restrain the CMB fibers from interacting and accumulating with $B_{\theta Gp}$ fields inside the baryonic gravity zone. These types of constraints, such as incompatible oscillation frequencies between fibers, lack of available disk space, etc..., were similarly examined for electron B_θ fields previously. Such constraints also ensure that, beyond the baryonic gravity zone, the only permitted interactions between the CMB fibers and the $B_{\theta Gp}$ field are those that extend the $B_{\theta Gp}$ field, rather than opposing and shortening it.

Galaxy or groups of galaxies that are locally but not gravitationally bound to other structures must have been created by gravitational processes that are independent of such structures. As the hydrogen, stars, and more complex objects comprising a galaxy or group of galaxies are being created, their overall principal gravity field must be created in the same proportion. During creation of individual structural parts, the corresponding developing gravity fields are nurtured by, and thus naturally attuned and synchronized with, the principal gravity field of the entire structure. In turn, each new component of the gravity field contributes to the principal field and becomes part of its sum. Two galactic structures within gravitational range may not gravitationally interact if their principal gravitational fields have developed independently. Unless their $B_{\theta G}$ fields were attuned and synchronized, the two structures would undertake a neutral relationship.

A black hole, suspected of existing at the Milky Way center, is under immense self-imposed gravitational pressure due to its great density and volume of matter (B_θ fields). Under such extreme pressure, the B_θ fields in its core are compressed to the extent that their individual fibers/disks are squeezed toward their origins. Consequently, the B_θ fields vanish (due to the pressure and crowding at the origins) as the fibers separate and are discharged from the black hole. Because of the extreme compression of the B_θ fields, all discharged fibers (photons) acquire the energy levels of x-ray and γ-ray fibers prior to their emission. In fact, projecting above and below the central galactic disk is an eruption bubble containing photons of these energies[9]. As the black hole draws in galactic matter (B_θ fields) in the plane of the galactic disk, it simultaneously discharges fibers (from the B_θ fields contained in the black hole) in the perpendicular direction. Thus, the Milky Way space is apparently being replenished with the fundamental heat fibers that were once its original building blocks, and will likely continue this role.

Notes & References:

1. The crossover zone (assigned as z = −1 fm to +1 fm throughout this study) is that section of the B-field where each half-field is influenced by the other. For example, the innermost disk of the right half-field begins to feel the effects of its left-hand counterpart at z = +1 fm during an inward oscillatory movement; the effects the disk experiences are sharp declines in v_z and B_θ in the range z = ±1 fm. In this study, the B_θ increments that constitute a half-field are summed in Equations (c′) and (c″) at z = 1 fm because the net B_θ intensity in the crossover zone is always less than at this point. Equations (c′) and (c″) are particular applications of Equation (c), and are reviewed later in Section 2. Owing to the opposing rotational and translational directions of the innermost fiber (or its B_θ disk) in each half-field, a relatively large radial force is imposed on each disk. This can be seen in Equation 1(b) for large $\partial B_\theta/\partial z$, where two B-disks rotate in opposite directions and ∂z is very small. Thus, the innermost disks of the half-fields have the smallest diameter (or shortest fiber oscillation stroke) among the disks in the electron field. The outward force E_z that mobilizes in the crossover zone as both half-fields contract arises from interaction between the two innermost disks. This can be seen in Equation 1(a), in which the $\Sigma h B_\theta$ increments are proportional to the ΣE_z increments and to v_z at the crossover zone. The $\Sigma h B_\theta$ increments are obtained for each half-field by summing from theoretical infinity to z = 0 (for example, between 1×10^{-16} and 1×10^{-18} m) by substituting appropriate z and v_z into Equations (c′) and (c″). The E_z force potential thus obtained from Equation 1(a) is expressed as N/C; this is converted to N per electron, which corresponds to a half-field, and finally to the E_z potential felt by the innermost disk. The latter value is obtained by multiplying N per electron by the ratio of B_θ of the innermost disk in the crossover zone to that at z = 1 fm using Equations (c′) and (c″). This conversion is appropriate because B_θ at z = 1 fm is the sum of all B_θ increments in the half-field, which drives the Coulomb potential for a half-field (N/e) as previously examined, while B_θ at z = 0 drives the outward force E_z in the crossover zone.

2. Mathematically, this is embodied in Equation 1(a). The changes in B_θ relative to changes in *h* are greater at the corner diagonal A-C than at any other corner orientation or within either of the adjacent B-fields outside of the overlapped corner. The orientation of the zone of highest changes in magnetic intensity controls the initial orientation of the combined z-axis, which is perpendicular to the diagonal plane A-C as required by Equation 1(a). Physically, this phenomenon is attributable to the vector summation mechanism, which draws other B-vector components from both fields to the diagonal plane A-C, where the B-vectors had initially coupled.

3. The two opposing factors force the trajectory of the electron field to curve downward but also oppose each other, causing ejection of one or more B-field disks (fibers known as radiation). This phenomenon has been observed and measured. As disks are ejected, the electron field ingests some of the applied field to create new disks/fibers, then re-emits and admits disks/fibers, in a repeating cycle.

4. In creating the proton, only two electrons were merged for illustrative purposes, but because the proton mass and thus its at-rest energy (proton or electron fibers should not be regarded as having mass) is roughly 1836 times that of the electron, approximately this number of electrons must

merge into a single proton. Once the first two electrons have merged, additional electrons (which are slightly out of phase with the embryonic proton) may merge with the pair under conditions similar to that illustrated in Figure 8(a). In addition, already merged combinations may merge again, as shown in Figure 8(b). See Sections 8 and 8D for discussions on mass and energy measurements of electrons and protons. Because a proton consists of roughly 1836 electrons, additional electrons attempting to merge should be restricted by lack of space; this resistance now exceeds the attractive force. *[The mechanism for proton development discussed here and in Section 4 is a potential source of protons. Other sources may have existed as well; for example, the nascent universe may have consisted of fibers joining in groups that rotated in both directions initially instead of only one. Nonetheless, the main emphasis being made is that protons are constructed similar to electrons, but contain more fibers that twirl in the opposite direction].* Because the electron and proton possess the same absolute charge (field strength in the context of this study), the effective B-field intensities and distributions of the proton and the electron must be the same. Because the proton contains more fibers than the electron, it presumably possesses greater B-field strength. However, due to space restrictions (*especially in the inner central area by the origin and z-axis where the B-strength is highest*), emerging perpendicular parts (B-elements) of the radial fibers are limited to the available space, which limits the strength of their B-fields. Hence, the electron field should be constructed to maximize the B-field strength of its individual fields. While overcrowding persists, any additional fibers attempting to join the ensemble will be rejected. Although they can reject individual fibers, electrons cannot reject the overwhelming force of attraction between their two full fields, as depicted in Figure 8(b), during proton creation. The density of fibers in the resulting field is increased but the field is less effective at optimizing its B-strength and distribution. As the space available for the perpendicular B components diminishes, each fiber contributes less to the B-strength. Overall, however, the B-field intensities and distributions are those of the electron. Portions of perpendicular B-parts prevented from developing in planes parallel to the x–y plane are accommodated in planes parallel to the h-z planes.

5. Here, the vector summation mechanism refers to the sum of the y-components of the two B_θ fields. In the r_x zone depicted in Figure 14, the y-components of both fields point in the same direction (upward in the figure), so the vector summation is additive. Outside the r_x zone, the y-components of the two fields orient in opposite directions, implying a subtractive vector summation. The overall net summation of both fields is additive and upward in the figure because the r_x zone contains the portions of both fields whose components are close to their origins, and therefore of higher y-component field strength. Outside the r_x zone, the strength of a portion of one field is greater while that of the other field is weakened by its greater distance from its origin. The net sum of both fields, and thus the Lorentz-type force between them, is proportional to the product of the y-components of the electron and proton B-field intensities, both measured at the proton field origin or *vice versa*, as illustrated in Sections 3C and 4 for similar interactions. The product of the B_y intensities reflects the vector summation mechanism of the y-tangential components of the two B_θ fields, which in turn generates the attractive force between them.

6. Equivalently, this mechanism can be viewed as the electron (proton) field bands of both bodies merging into composite electron (proton) field bands. As the composite proton field bands pass the composite electron field bands, the gravitational force is mobilized.

7A. Although "heat (photon) fibers" possess no intrinsic heat, they are referred to as "heat fibers" in this study because they give rise to our standard notions of heat, as subsequently explored in Section 11C. The true substance of a "heat fiber" is energy. The fiber source and its material and energy sources are unknown, as they are for the conventional concept of the electron or any other particle; thus substantiation of either must be inferred from appropriate mathematics derived from observations and tests. In this study, the appropriate mathematics is Maxwell's equations and their extension. In Section 1 and Figure 1, fibers are regarded as fluidic, possibly mimicking electric arcs, but are many orders of magnitude smaller, of course. Nonetheless, the content of "heat fibers" is energy rather than heat. Individual fibers cannot be visualized and are examined as photons for the first time in Section 6.

7B. As examined further under *Temperature* in Section 11C, temperature measurements reflect the density and energy of free fibers in contact with a measuring device, not our notion of hot and cold, which have no scientific meaning. Thus, heat and temperature need not be associated with individual fibers. Fibers (identified as photons for the first time in Section 6) are referred to as "heat fibers" because they directly affect our notion of hot and cold. For example, if one touches an object that transfers radiation (heat fibers) to the fingers, the individual senses that the object is warm or hot; conversely, if the fingers transmit radiation (heat fibers) to the object, the individual would sense that the object is cool or cold. In both cases, the object itself is neither hot nor cold.

8. In addition to free heat fibers, free groupings of fields that include heat fibers can occur in special cases, which impact temperature measurements. These cases are not considered here because the vast majority of temperature measurements are affected by free heat fibers only.

9. Dennis Overbye: *Bubbles of Energy Are Found in Galaxy*. The New York Times, November 9, 2010

10. For larger velocities, see "Biot-Savart law for a point charge" originally derived by Oliver Heaviside in 1888, which is adjusted for relativistic effects.

11. This is appropriate because the B-elements occur individually (not continuously) in the disks. Thus, the elements responsible for the highest attraction (or opposition) are those whose tangents are perpendicular to the y-axis. It is this force that is measured.

12. If an electron and proton fields are oriented as shown in Figure 14, the B-vectors in the r_x zone are amicable, and their summed half-fields yield an attractive F_x force. Although this Lorentz force (which far exceeds the force of gravity; see Section 4) develops between the electron and proton, the net field is zero because the half-fields of the electron or proton cancel beyond the locality of the interaction. Thus, the oscillation of electron or proton half-fields along their z-axes cannot generate a gravitational field.

13. In general, Einstein's general theory of relativity includes relativistic adjustments to Newton's field equation for gravity. The strengths (or intensities), distribution and orientations of the $B_{\theta G}$ field bands within each and between two gravitationally interacting bodies behave

relativistically (per general relativity) under appropriate conditions. For example, an interacting body, translating at high velocity relative to its partner, will have field bands that are adjusted in intensity, distribution, and orientation relative to its partner. In another example, if the second body is extremely massive, the intensities, distribution and orientations of the field bands of the subject body are measurably adjusted also, because the fibers contained in the subject body's B_θ fields are affected by the large gravitational $B_{\theta G}$ intensities of the second body. These internal adjustments in turn affect the gravity field (which is derived from the B_θ fields) of the subject body.

14. As presented in this book (for example in footnote 1 of Figure 2 and elsewhere), the perpendicular components that form along the radial fibers constitute the circumferential magnetic B-field of the electron. The circumferential B-field intensity is measured in Teslas, or Webers per unit area, in the tangential direction of the B-field. The B-field intensity or magnitude depends on the density of the perpendicular components *(which is enhanced near the z-axis, where the volume of perpendicular parts is greater, Consequently, B increases in the central regions and decreases further away, where the volume occupied by perpendicular components is less)*. The B-field intensity also depends on the tangential speed *(which increases further from the z-axis, thereby increasing B)*. These contradictory actions exert opposite effects on the B-intensity. The region of maximum B-intensity (whose magnitude is given by Equation c′) occurs between the z-axis and the outer edge of the B-field. The tangential speeds of the perpendicular components result from the angular velocity of the twirling fibers illustrated in Figures 2, 3 and elsewhere. *(The region of maximum B-intensity was calculated as h = 0.707z).*

15. Discussion on oscillating vectors as a means of comprehending the congregating behavior of fibers in the first two paragraphs of Section 2 is only a simple introduction to how the fibers may join and congregate via what is called the "vector summation mechanism". This introduced mechanism is similar to the actual vector summation mechanism that is defined subsequently in Section 2. This introduction is necessary to set up the presentation of the electron field upon which Maxwell's equations and its derived extension could be applied to demonstrate how fibers do actually join via the actual "vector summation mechanism", and whole fields can be created as a result. Reference to further examination of the vector summation mechanism, subsequently in Section 2, is made in both introductory paragraphs of Section 2.

The substantiation of the vector summation mechanism starts in earnest with the introduction of Maxwell's Equations 1 and 2 and in particular Equations 1(a) and 1(b), all in Section 2. *(It is assumed that the reader is familiar in the utilization of Maxwell's equations and the following observations are self-explanatory).* Due to the orthogonal nature of the curl equations of Maxwell, a requirement is established by such equations, that the field heat (photon) fibers be oriented parallel to each other and perpendicular to their B-elements, and move orthogonally to both orientations. Per Equation 1(a) for example, the field fibers move in the direction of the E_z type force, which is orthogonal to both the fibers and their perpendicular B-elements. Per Equation 1(b), the field fibers contract or expand in the radial direction due to the E_h–type force, while they move in the direction orthogonal to E_h, themselves, and their perpendicular B-elements. Physically, the mechanism in both examples occur as fibers are drawn *(or repelled if directions of movements of B-elements are opposed)* to the perpendicular B-elements of nearby fibers in conjunction with relative motion between the fibers that is perpendicular to both the fibers and their B-elements. For lack of a better term, this process is the "vector summation mechanism" that is alluded to throughout the book. Equations 1(a) and 1(b) dictate the parallel orientation of the planes in which, the individual fibers oscillate and twirl, and it also represents the mechanism that gives rise to the oscillation of the two half B-fields to and fro along the z-axis of an electron, for example, per Figures 2 and 3. Equation 1(b) also dictates that the fiber origin shown in Figure 1 for all fibers is on the z-axis shown in Figure 2. Hence, the movements and adjustments that fibers make as they congregate and interact[23] *(as suggested in the first two introductory paragraphs of Section 2)* and form B-fields, such as the electron in Figure 2, are governed by the mechanism depicted in Maxwell Equations 1 and 2 and its derived extension, which gives the oscillation velocity of the half B-fields. More detail on this subject including the formation of B-fields is provided in the rest of Section 2 and in Section 11.D.1. An example giving the interaction of two fibers (photons) using Maxwell's Equation 1(b) is presented in Section 11.D.2.

16. Linnyk O, Bratkovskaya EL, Cassing W. (2015) "Effective QCD and transport description of dilepton and photon production in heavy-ion collisions and elementary processes". Prog. Part. Nucl. Phys. 87: 50–115

17. Edwards, D. (1981). "The Mathematical Foundations of Quantum Field Theory: Fermions, Gauge Fields, and Super-symmetry, Part I: Lattice Field Theories". Int. J. Theor. Phys. 20(7): 503–517.

18. Birkhoff G, von Neumann J. (1936). "The Logic of Quantum Mechanics". Ann. Math. 37 (4): 823–843.

19. Brown LS. (1994). Quantum Field Theory. Cambridge University Press. p 400.

20. A similar derivation of the electron model that builds upon Biot-Savart's formulation for the cylindrical B-field of a translating point charge can be found here:
Correnti DS. (2018). "Mechanisms explaining Coulomb's electric force & Lorentz's magnetic force from a classical perspective". Elsevier's *Results in Physics*. Volume 9, June 2018. 832-841. Link address: https://www.sciencedirect.com/science/article/pii/S2211379717325871

21. Atkins PW. (1974). Quanta: A Handbook of Concepts. Oxford University Press. p. 52.

22. Correnti DS. (2018). PROJECT: Hadronization of Scattered Particles in High-Energy Impacts. ReseachGate. https:// www.researchgate.net/project/Hadronization-of-Scattered-Particles-in-High-Energy-Impacts.

23. Masterson, Andrew. (2018). Bonded photons represent a new form of matter. Cosmos. https://cosmosmagazine.com/physics/bonded-photons-represent-a-new-form-of-matter
The photonic matter discovered in this research would also occur in other environments where there are extremely dense mixtures of them such as occurs in this referenced experiment.

24. Although the proton field has the same magnetic field strength as the electron field, its photon fibers are much more energetic. Although photon fibers have no 'mass', which is a property but not a physical object in this stdy; the proton mass is much greater than the electron mass, because its energy is much greater. For example, a much greater force is required to cause a given acceleration of a proton than an electron because the proton's internal energy is much greater than the electron's energy; it takes more force to move more internal energy (thus more mass). For example, the property of mass is created due to such internal energy (photons) interacting with each other in a group [23] such as a proton field. The greater the photonic energy, the stronger its interactions are, which gives a greater mass of the group. Although an individual photon has energy, it has no mass because it is not interacting with other photons in a group such as a proton field.

www.ingramcontent.com/pod-product-compliance
Lightning Source LLC
Chambersburg PA
CBHW081606170526
45166CB00009B/2845